The Age

of the

Artificial Person

Alastair Waterman

The Age of the Artificial Person

Contents

Introduction

Using AI is very convenient for us. In just a few short years, as AI has got better at doing things, the companies behind these technologies have become incredibly successful. We cannot expect that people will halt the development of culture for the first time in their history and say 'stop' to technology. This process is completely inevitable and cannot be suspended for any significant length of time. The benefits of AI are so obvious that any attempted bans will lead to nothing. It is necessary to understand that this is not accidental and that the development of AI most naturally comes from our desire to gain an advantage over nature – to be different from nature. And we want AI to help change us, too.

In our time, the question of human nature takes on a new meaning. Previously, we tried to understand how we differ from animals, but now we are faced with a more difficult problem – to distinguish ourselves from artificial intelligence, which is becoming smarter every minute. How can we do this if AI imitates us so skilfully and already does many things better than us? In *The Age of the Artificial Person,* this difficult issue is explored. In the book, we try to understand the actual nature of a biological person, and conclude that a person is distinguished by a specifically human way of connecting with reality.

Old words such as reality, feeling, consciousness and culture acquire new meanings in this book. For example, artificial intelligence has incredible computing abilities, but it still has no connection with reality, even if equipped with video cameras and other sensors. The basic form of connection with reality that is available to animals, namely *feelings*, arose as a result of billions of years of evolution. Unlike Chat GPT, the most ordinary frog is more connected to reality and knows more about the real world than any modern artificial intelligence is capable of.

Evolution has also created a more complex way of communicating with reality – consciousness. Consciousness differs from the ability to feel in that it forms and gives meaning to surrounding objects and processes. And this meaning differs from the one that was laid down by nature. This is how a culture arises, which is characteristic not only of humans because bees also create their own amazingly complex hives. But humanity has made culture our tool for transforming the world around us and, of course, most importantly, a tool for changing ourselves. The material and immaterial culture that we have invented has changed us beyond recognition and continues to change us every day. And the development of artificial intelligence is another step that our culture is taking to change us again.

The Age of the Artificial Person is an analysis of why and how humans will change and how we will gradually become artificial, but also a manifesto on why this is possible, why it is not dangerous for humanity, and why it may even be inevitable. The book provides discussions about what consciousness is, and arguments in favour of why it is possible to create a conscious machine. It also explores the question of what components the human personality consists of and how to transfer the personality of a biological person to an artificial medium.

The era of the artificial person is a natural step in evolution. We are accustomed to perceiving ourselves as we are today, but our modern image is only a captured moment of human evolution. In fact, we, as the driving force of development in the universe, begin in the primordial biological soup and will continue in the form of artificial beings, completely created as we wish. But what does this primordial soup, the animal world, artificial intelligence, and us all have in common? This is probably the impulse of life that humanity was able to transform into the impulse for the development of culture in the broadest sense, as something different from nature. The book explores the question of how exactly humans use culture to gain advantages over nature and differentiate themselves from nature. Man creates his own way of accumulating useful changes in the form of material and intangible culture. This method differs from other forms of accumulation of useful changes: genetic, social, etc. In the book, five stages of this process are described – five eras five ages.

In the **Age of Coercion**, a person goes beyond the instincts of a herd animal and forms an institution of management – leadership. This leadership initially consists of various forms of coercion but gradually crystallises in the cultural institution of folk traditions, and the class of aristocracy also emerges.

In the **Age of Persuasion**, a person creates a new, more complex form of management – through the formation of common group and public values. A convinced person no longer needs to be forced. In this era, a class of clergy arises in a broad sense, which includes all forms of activity involved in the process of forming values – religion, art and philosophy.

In the **Age of Enlightenment**, the form of control over human forces becomes even more complex. Mass education leads to the fact that common values no longer need to be conveyed in the form of persuasion since a person acquires values when receiving an education.

In the **Age of Inspiration**, the final stages of which we are witnessing today, the form of control of popular forces reaches its highest development – within the framework of the cultural institution of ideology. Ideology, in one form or another, has turned out to be necessary in order to successfully manage not only educated but also uneducated people, giving the latter, through ideology, access to necessary knowledge in the simplest and most easily memorised form.

The peculiarity of the **Age of Artificial Person**, which could also be called the **Age of Artificial Skills**, is that thanks to artificial intelligence technologies, culture will soon reach that fundamentally new level when the need for control by conventional means through coercion, persuasion, enlightenment or inspiration will no longer exist. Artificial intelligence is the first economic agent in our history that does work without outside pressure.

How can AI change us? Beauty is a form of efficiency, and by being potentially more efficient than us, AI can open up new horizons of beauty for us, helping us see the world as even more complex and beautiful. The idea of "**Dark Science**," or a science that will be completely devoid of the influence of human subjectivity and anthropocentrism will be discussed. Today, AI already solves many theoretical problems in mathematics, but it can also succeed in physics, biology, medicine, astronomy, etc.

But this is not the main thing. The most important thing is that AI can change us in at least two directions. Firstly, it will be able to expand our senses and teach us to better understand the universe, to see meaning where we cannot understand it because of our anthropocentrism. It will teach us to understand patterns in much more complex systems than our minds can handle. One way or another, the ability to understand something depends on how much independent data the agent is able to compare, and AI is significantly superior to us in this indicator. Potentially, AI will be able to discover beauty for us where we missed it or did not notice it and be able to comprehend new forms of art.

The second thing that AI can give us is the eternal life of an individual in an artificial body. Today we can train AI to answer questions in the same way as a specific person would. That is, we can actually make an intellectual copy, a cast or a meaning graph of a personality. But it will only be a dead copy of the mind and not the personality itself. Further research into AI, which we can call **artificial personality technology**, would make it possible to combine a person with AI and ultimately place that personality in an artificial medium. And this, too, apparently, is our inevitable future. This will bring many problems to humanity, but will also herald a completely new historical period in our culture. This will be the culture of a person who has left the cradle of his body.

About the Artificial Person

Can Robots Feel? Multiple Realisability or Special Environment

Conversations about whether machines can theoretically feel have been going on for quite some time. But it was only recently that this became a pressing issue when Blake Lemoine, an engineer at Google, stated that the LaMDA (Language Model for Dialogue Applications) system has feelings equivalent to the level of an ordinary child. ('The Google engineer who thinks the company's AI has come to life, *The Washington Post*, 2022) The company denied this statement, and in general the scientific community received this news without much enthusiasm, but formally, the discussion about the feelings of machines reached a fundamentally new level. Now, this is not just a philosophical question.

Strictly speaking, neither the Google engineer's statement nor the company's denial of it can yet be recognised as true, if only because we do not have reliable criteria and an accurate definition for such a phenomenon as "feeling." At the level of general ideas, we can say that we are talking about the fact that Blake Lemoine discovered similarities in the language behaviour of the LaMDA system with the way children behave in similar situations. And by analogy, he concluded that external manifestation (speech or text) must also correspond to internal content (feelings, consciousness, inner world). But the problem is that we do not know at all what feelings, consciousness and inner world are.

By analogy, we think that the "germ" of the inner world in nature is the animal's ability to experience the simplest feelings, such as pain or irritation. It seems to us that this gradually evolved into the "rich" inner world we know. Therefore, we are inclined to believe that all the properties of the inner world are, in one way or another, connected with some sensitivity of animals. This means that if robots are not living beings, then it is as if they cannot feel. But how can you check this?

The problem is that without a definition of what a feeling is, we cannot design an experiment to test it. On the other hand, functionalists have proposed a theoretical concept that aims to refute our initial intuitions. This is the concept of multiple realisability (Putnam, 1960). In accordance with it, the same functional state can theoretically be realised in different ways. That is, if, for example, there is such a functional state as "pain," then it can be realised not only in the animal world but also, for instance, in technology at the level of artificial machines.

To some extent, the concept of multiple realisability is confirmed by various examples. For example, we know how to create robots that can perform movements. The ability to make movements is inherent in animals and, in general, is initially of biological origin, but the robots that we create are not biological, although they can make movements. Therefore, it is clear that the ability to make movements is not unique to life and can be realised in various ways. Such an ability is not a reliable sign that this is an animal, much less that it has feelings and an inner world.

Another important function of life is the ability to adapt, and part of this function is the ability to exercise intelligence and solve various problems. This ability was realised by nature in a biological medium, with the help of the nervous system of animals. But then man created computing machines, computers that today can solve many problems just like humans. The level of problems solved by computers is very high, and we do not yet know where the limit of the "intellectual abilities" of the machine is. Thus, the ability to solve problems is not unique to biology, and if we have an object in front of us that can solve problems, then we cannot be sure that it is a person.

Formally, machines have not yet passed the Turing Test with complete reliability. However, their ability to solve problems and imitate human behaviour is impressive. This is proven by the example of the Google engineer given above. This generally means that passing the Turing Test is also no longer a strict proof that this is a person and that he or she has an inner world. Rather, we could say that there are "biological intelligent agents" such as humans and artificial intelligent agents such as LaMDA or ChatGPT.

But making movements, solving problems, and feeling are fundamentally different things. It seems that problem-solving is a much more complex phenomenon than movement, as is the ability to have a feeling of the fundamental need to solve problems. These three functional states seem to differ in their basic nature. But are they really fundamentally different? After all, the first two of them confirm the principle of multiple realisability and can be implemented in artificial technology, but what about the last one? Is there some impassable barrier to a robot starting to "feel"? Or will the same thing happen to this ability that has already happened to the ability to move and solve problems?

If technology has come so far, why shouldn't it take another step and begin to feel? But in such matters, extreme precision is required. Maybe we are comparing apples and oranges here. Let's try to figure it out.

Movement is always a transformation of energy. In a narrow sense, any movement is associated with the transformation of energy, and in a broad sense, movement *is* the transformation of energy. For our purposes, however, it is enough to know that both living organisms and artificial mechanisms make movements due to the transformation of energy. Actually, an "engine" is a certain organ or device that converts some type of energy into movement. To be able to move, living organisms consume energy in the form of food, water, various substances, solar or thermal energy, inhaling oxygen, and so on. This energy is then converted by the animal's organs into kinetic energy or movement. Artificial mechanisms also convert fuel energy into kinetic energy.

Movement can be realised in various media, both in living organisms and in artificial machines, insofar as both energy transformations can be carried out. And there is no doubt that the movement in both cases is of the same fundamental nature. This is the transition of chemical, electrical or other energy into mechanical energy. That is, with movement in terms of multiple implementations, everything is simple.

The ability to solve problems is a more complex phenomenon, but not so complex that we do not understand its basic principle. First, let's give an example. To catch up with the antelope, the cheetah chooses the shortest trajectory to its target. Calculating this trajectory is not an easy task, but the cheetah copes with it. The process of hunting, in this case, consists of two fundamental moments, one of which is the actual rapid movement of the cheetah towards the antelope, and the second is the choice of the trajectory of this movement. For it to work, the predator needs enough energy, which it spends on this manoeuvre.

Without sufficient energy expended by the animal's muscles, the hunt would be unsuccessful. However, the choice of trajectory, for which the animal's nervous system is responsible, is also no less important. The animal's brain also consumes a lot of energy to calculate the running trajectory. However, the essence of what the cheetah's brain does in this case is not in transforming energy but in carrying out calculations. That is, by and large, the solution to the problem here is calculation or, in other words, performing operations on information. By analogy, we can talk about the transformation of information.

The brain solves problems by transforming information. These transformations by themselves do not force the cheetah to move since movement also requires muscle strength, but at the same time, the animal's nervous system directs these forces in the right direction. In fact, solving any problem means carrying out transformations with information to effectively distribute resources (energy). The brain chooses where the cheetah runs, and the muscles put this decision into practice.

The same principle of separation of two functions is also implemented in artificial machines. The bulk of the machine's energy is spent on movement, while operations to perform trajectory calculations most often require less power. In this sense, the "problem-solving" function is presented as informational, not energetic. Solving the problem here is like turning a steering wheel, which in itself does not require much energy but leads to the fact that a heavy car changes its trajectory. But another question arises: in which direction should the steering wheel be turned?

Since we are doing philosophy here and not ballistics, we need to understand what solving a problem is in some of the most fundamental senses. For example, language models such as ChatGPT can solve problems, and they also carry out information transformations. But ChatGPT, just like humans, makes mistakes. In this sense, solving a problem does not mean solving it correctly. In a sense, any transformation of information is a solution to some problem, but in relation to each specific problem, it can be correct or incorrect.

For example, if we ask a four-year-old child what two plus two is and he answers five, then this situation will be completely different from another in which he answers, say, one. In the first case, we will understand that the child has an idea of addition but made a mistake. In the second, his solution will show us a

17

lack of understanding about addition. If he answers "bunny" to this, then we will understand that he most likely has no idea about numbers at all. But at the same time, each of the child's answer options will be the result of information transformation, and each will also be a solution, with varying degrees of effectiveness in relation to a specific task. Let us note here that efficiency is one of the fundamental characteristics of solving a problem.

To describe what a solution to a problem is as a fundamental phenomenon, we therefore need to divide it into two sub-phenomena or even into two dimensions. One dimension is information transformation, which provides an answer to how resources should be allocated. And we will say that any such transformation will be a solution to some problem. And the second dimension is an assessment of how effective, expedient, and so on, this solution is. If we do this and divide the whole process into these two elements, then it becomes clear how exactly animals use their ability to "feel".

The fact is that feeling turns out to be precisely the natural mechanism that allows an animal to evaluate how correctly it solved a particular problem. For example, if a cheetah catches up with an antelope and enjoys a meal, then after the fact, it concludes that the trajectory was quite effective. This process is called positive reinforcement, and it is part of the process by which the cheetah's neural network learns how to hunt antelope. In this case, the cheetah's teachers are also his older relatives, but mainly these are his own feelings. They tell the cheetah when there is success and that this should be repeated, and when there is failure and such decisions should be avoided.

Artificial intelligence systems have a similar process, which is called supervised learning. Such language models as LaMDA or ChatGPT are trained by adjusting the coefficients of the neural network, and so far, only we, that is, people, are the teachers for them. We use our senses, as well as a more advanced decision-making tool called consciousness, to train these AI systems. That is, problem-solving for artificial intelligence systems is divided into two parts. The information transformations themselves are performed by a machine, multiplying large multidimensional data matrices by each other, and we, those with the ability to feel, still evaluate the effectiveness of these decisions.

In this sense, we seem to have narrowed the question somewhat: can a robot feel? Formally, we can already say that he cannot feel, at least until he is able to assess the correctness of his decisions. The judge of the performance of robots' computing activities is still a person: either a code developer, a teacher in machine learning, a consumer giving feedback to developers, or a legislator etc.

But what is the ability to evaluate the correctness of decisions? What kind of phenomenon or transformation is behind this? If we understand this, we can also understand whether this functionality is subject to the principle of multiple realisability.

Indeed, we were able to implement the functionality of "movement" in machines because it turned out to be a phenomenon that was understandable to us – the transformation of energy. And we know how to transform energy. And we were able to realise the ability to "solve problems" in robots, because this also turned out to be a phenomenon that we understand – the transformation of information. And we also know how to do this. But what is the ability to evaluate the correctness of problem-solving? What is this skill in some fundamental sense?

Surprisingly, but necessarily, the ability to evaluate the correctness of problem-solving is what connects us with reality. We can solve the same problem in a million ways and get a million different answers, and only a connection with reality will force us to choose one of the solutions as suitable. Nothing other than our feelings performs this task. It is the feelings that give access to what is really happening. This does not mean at all that the senses never deceive us and that various illusions are not possible, but it does mean that without the ability to feel, with only the use of reason, we would never be able to distinguish reality from a dream or a figment of our imagination.

In this sense, a language model such as ChatGPT cannot itself evaluate how relevant the text it produces is to reality. To do this, it would have to have access to reality. But what is access to reality? In AI terms, we would say that such access would be the ability of an AI to go beyond what it has been taught on its own and see how things are independent of that training. This is how a person with feelings acts.

Let's give an example. The Checker Shadow Illusion is well known.

Pic. 1. Checker Shadow Illusion

In this illusion, cells A and B are the same colour, but due to the settings of our brain (prior conscious experience), it seems to us that B is much lighter. It would be a mistake to say that our feelings let us down in this case. It is not our feelings that fail us but our higher levels of perception. For our sensations, cells A and B are the same, but because consciousness is based on some previous experience associated with the perception of light and shadow, we are aware of A and B differently. However, we can go beyond what our brain's neural network is trained to do and see what the colour of a cell is. After all, we have this amazing connection with reality.

There are no illusions for AI because reality itself does not yet exist for it. No matter how much we would like to believe it, robots do not yet possess anything that would allow them to distinguish between these degrees of being. What is available to AI is not reality but information about reality, and we have yet to understand how these things are fundamentally different from one another. The video camera tells the AI data about reality but does not give access to reality itself.

The need for sensory or empirical knowledge of the surrounding world was noted by empiricists, for example, Francis Bacon. But the Greeks thought about this long before Bacon. For example, Plato gave the metaphor of prisoners chained in a cave, who observed only the shadows of people passing by the cave falling on the cave wall. But they didn't see the people themselves. In other words, even the ancient Greeks discovered how important it is in cognition to separate the things themselves from how they appear to us in phenomena. To get from phenomena to things, one must at least distinguish between these two categories. There is no such division for AI.

So, we believe that the ability to evaluate the correctness of problem-solving is fundamentally nothing more than a connection with reality. This is not a transformation of information and not a transformation of energy. Perhaps this is not a transformation at all, but something different in nature, say, a way of existence or a special environment. Frankly, we don't know what it is, but this type of phenomenon, this functionality, is so different from the ability to move or solve problems that we cannot, by analogy, attribute this ability to robots. There is no reason for this yet.

Biological Consciousness and Artificial Consciousness

What is consciousness? It's amazing how many arrows are broken by this problem. A good half of all philosophy, in one way or another, directly or indirectly, is devoted to this issue. In this book, we cannot avoid this problem either since an artificial person is a conscious being created artificially. Therefore, when discussing artificial man, we are obliged to at least outline how, in our opinion, this task of millennia should be solved.

We will not present all the conclusions and the course of reasoning, but perhaps we will limit ourselves to one idea that looks like the most important. Consciousness is a multi-layered and multi-level phenomenon, which is what it is only when all layers and levels are involved.

Consciousness is a complex physical-bio-psycho-socio-cultural phenomenon, and to understand it, it is necessary to find a "particle" of a physical-bio-psycho-socio-cultural nature without "reducing" it to any of these poles. Researchers usually focus on one or two components of consciousness, but this becomes an obstacle to understanding this phenomenon.

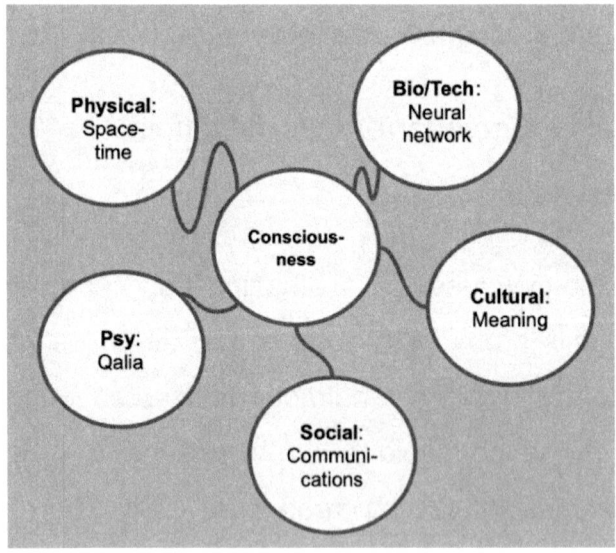

Pic. 2. The nature of consciousness.

Theories of consciousness that focus on the brain and the functioning of neural connections are primarily the biological nature of consciousness. Such a concept is, for example, Baars' theory of global workspace. According to Baars (Baars,1988), consciousness is a collective action of parts of the brain, where different departments work together to solve complex problems, creating a kind of "workspace" for wider access to neural resources and memory. Another important theoretical approach to the neurobiology of consciousness can also be found in these authors. (Crick & Koch,1990).

Theories of consciousness that consider consciousness as a special form of physical state are physical theories, which include, for example, the quantum theory of consciousness of Penrose (Penrose, 1987), who believes that consciousness is a quantum effect. He believes that quantum processes take place at the level of the microstructure of neurons, which form a special subjective experience.

A mixed physical-biological theory can be called Tononi's concept of integrated information. (Tononi et al., 1992) and (Edelman & Tononi, 2000). It is mixed because it uses information as a key phenomenon, considering it at the intersection of physics and biology. We also include the naturalistic dualism of Chalmers, who believes that consciousness is a special type of immaterial information, as a physical-biological theory. At the same time, he admits that consciousness is as physically fundamental as, for example, an electric charge or space. Although Chalmers calls consciousness "non-physical," it becomes clear from his comparisons that he is simply talking about still-unknown areas of physics. (Chalmers, 1996).

Without the physical component, a theory of consciousness would be incomplete since we see that our "perception" is very different from everything else that we observe in the world. Consequently, if our perception somehow physically exists, then it exists differently from the way things and processes known to us exist. Without the physical part of the theory, we do not understand how and where the mental "space" of our consciousness exists and what fills it. All arguments against physicalism claim that physics is insufficient to explain consciousness, but, in fact, they prove our inadequate knowledge of space and the topology of the world. (Nagel, 1974).

But physics alone is not enough, and the absence of a consideration of biology for the theory of consciousness is unacceptable, since the participation of neurons in the formation of consciousness is also undeniable. Conscious processes occur together with some neural processes and are impossible without the latter.

The problem with purely biological theories, however, is that they do not consider the indisputable fact that consciousness in some way depends on the upbringing mental, social and cultural experience of a person. For example, if you show notes to a person who does not know musical notation, he will not be able to "see" the sounds behind the notes, while an experienced musician, on the contrary, will not be able to help but see how these notes sound. What a person who does not know a language cannot see in a text or hear in speech, a native speaker again cannot help but see and hear.

To this, a neuroscientist will say that a person's experience and skills are recorded in his neural connections, including, say, knowledge of a particular language, and nothing more than a consideration of neural connections is needed to understand the phenomenon of consciousness. To this, we will answer that neural connections are just "letters" of consciousness and not language at all. To understand the "language" of consciousness, in addition to letters, we need rules of syntax and semantics. And suppose the "syntax" of consciousness can still be associated with neural processes. In that case, the rules of the "semantics" of consciousness are formed outside the brain and are associated with external objects and processes. Meaningful connections between things are formed mentally, socially or culturally. For example, if there is no word, then it is very difficult to think about the concept that this word expresses, and it is even more difficult to convey the thought about it to another.

The psychology of consciousness as such can be considered a mental theory of consciousness, since it considers the mental conditions under which objects are perceived consciously. This is, for example, Wundt's theory of attention (Wundt, 1912), in which consciousness turns out to be the focus of attention. This creates an understanding of the energetic barrier or level of consciousness that the stimulus must overcome to be conscious. In this sense, psychoanalysis became a more complex psycho-socio-cultural theory of consciousness, as it linked the energy of the stimulus with social (Freud) and cultural (Jung) factors.

According to psychoanalysis, some content may not enter consciousness, even if it has great energy, but is blocked for social or cultural reasons and remains in the unconscious. Psychoanalysis was an important step towards the formation of a theory of consciousness. However, it missed the biological and physical aspects of this phenomenon. Because of this, psychoanalysis explains only "pathological" deviations from the norm but not the phenomenon itself as such.

Psycho-socio-cultural theories can also include the concept of Vygotsky (Vygotsky, 1962) and Soviet Marxist psychology and philosophy in general. This direction turned out to be very important in understanding the connection between consciousness and personality, but it too ignored the physical and biological aspects of the problem, explaining only the transitional stages during training and upbringing, crises of personality formation. But a person realises something according to a complex physical-bio-psycho-socio-cultural mechanism not only when he is learning but in general at almost every moment in time.

The dispute between two Soviet philosophers of consciousness, Ilyenkov and Dubrovsky (Oittinen & Levant, 2013), was precisely devoted to this aspect of the theory. According to Ilyenkov, the socio-cultural aspect is completely autonomous and sufficient in explaining the phenomenon of consciousness, to which Dubrovsky wisely responded that with such a formulation of the question, the individual subject, the conscious agent, that is (we will say) the physical and biological part of the theory, is blurred. The role of language and the understanding of what meaning is was also explored by Searle (Searle, 1992). The concept of consciousness cannot be complete without taking into account the role of language and speech.

A mixed bio-psychic theory of consciousness can be considered the concept of neuroscientist Damasio (Damasio, 1989), who actively promotes the idea of a connection between consciousness and feelings and emotions in general. The problem with this theory, in turn, is that it does not consider the fundamental physical aspect and the cultural nature of consciousness. Damasio calls consciousness "the feeling of what happens" (1999), which ignores the cultural nature of the perception of "what is happening" as a construction of reality. Indeed, by "what is happening," Damasio means some external reality, essentially confusing the phenomenon of feeling and the phenomenon of consciousness, mixing them up. Damasio rightly defends the role of feelings in the formation of consciousness. However, combining two different processes into one is unacceptable.

So, there are many different theories, but they all reduce to one conclusion: the "Great Unification" is not happening yet. In defence of these and many more unlisted theories of consciousness, we note that for physical, biological, mental, social and cultural processes, it is very difficult, in principle, to find common ground.

In search of a unifying phenomenon, many researchers have concluded that it is necessary to use the concept of information. However, "information" in physics is a quantitative characteristic. In biology, it is reduced to the concept of code and code dependencies. In psychology, it is reduced to the concept of an energy barrier, and in the social and cultural spheres, information can only be considered from the point of view of the content of information. In other words, in all these environments, the concept of information is fundamentally different, and we are talking about synonyms rather than a single phenomenon.

Another important point in the formation of a theory of consciousness is to consider the emergence of properties of a certain type. Let us explain this as follows. If consciousness is a physical-bio-psycho-socio-cultural phenomenon, then evolutionarily, it could be formed in two variants. The first option – not emergent – would be that over time a certain physical phenomenon (let's call it f-consciousness) would be supplemented by a biological component (b-consciousness), and then supplemented by a mental component (p-consciousness), then a social component (s-consciousness) and, finally, the cultural component (k-consciousness). This is, for example, the Marxist approach (Yasnitsky, 2021) of Rubinstein's theory of reflection (reflection is a fundamental property of matter) or, for example, Tononi's panpsychism (everything has a psyche to some extent), as well as the largely naturalistic dualism of Chalmers (consciousness is a fundamental property like charge or space).

For these three theories, consciousness is like an "onion" consisting of layers of different forms of being. And if, for example, you remove the cultural component, then the remaining layers will still be preserved. The emergent version looks different, and it implies that all layers of consciousness, from the physical to the cultural, are interconnected, and the elimination of one leads to the elimination of all others. This means that, for example, the physical properties of consciousness depend on its cultural component.

There are cases where children lose hearing and vision at an early age. For example, the Soviet psychologist Meshcheryakov (Meshcheryakov, 1979) was involved in teaching deaf-blind children. He managed to preserve the consciousness of these children only thanks to the special practice of introducing the world "through his finger." Tactile sensations in the fingers made it possible to maintain socio-cultural communication in the child and, therefore, consciousness. Here, we see an example of the role that the cultural component of consciousness plays in maintaining its physical component (internal mental space).

We take the position that consciousness is emergent in the special sense that the phenomenon of "consciousness" manifests itself as the interaction of all these forms of being.

This means that we proceed from the position that evolutionarily, this phenomenon could develop not as the accumulation of layers on an "onion" and the transition from rudimentary consciousness gradually through the stages of "awareness" but as the emergence of new properties in different combinations.

Some form of sensuality in living organisms, for example, could appear on a biological substrate without additional physical or mental effects. Then, during evolution, this sensuality could develop in the social systems of organisms. Further complications could lead to the emergence of the simplest mental activity, which is then accompanied by the formation of rudimentary forms of the inner world or internal mental space. And finally, further complexity in the plane of socio-cultural development could lead to the expansion and enrichment of the content of mental space to its modern forms.

So, we combine holistic and emergent approaches to defining consciousness. Being holistic, such a definition would include all levels of consciousness, from the physical to the cultural, without giving priority to any of them. Emergence here could be reflected in the fact that all these levels are mutually dependent on each other as form and content.

Consciousness is an active spatiotemporal environment, formed by neural processes in the brain and forming feelings and personal subjective experience, which consists of active interaction in the social environment and with oneself through the language of cultural meaning units.

In other words, consciousness includes the following components: physical – mental space-time, biological – information environment, mental – sensory personality, social – communication, cultural – meaning.

The advantage of such a definition is that it considers all aspects of the phenomenon being described, but its weakness is the absence of any single basic concept that would permeate all forms of being and shed light on the integral nature of consciousness, given in the unity of all its forms.

In this book, we discuss what an approach to a theory of consciousness might look like that uses an integral concept of the problem. Even though the "problem" is traditionally associated with the subject (without the subject, there can be no problem), it can also be considered within the framework of the "dark approach", that is, before the subject. In some fundamental sense, a problem is a deviation from harmony or, in an even more general sense, a deviation from symmetry. Various forms of asymmetry are theoretically conceivable for all environments and at all levels of existence. The solution to the "problem", in this case, will be a process aimed at restoring symmetry in one or another of its manifestations.

The extension of the concept of "problem" to the inanimate world is not the humanisation of the world, but, on the contrary, it is the objectification of the human. Even so, the creation of artificial consciousness looks like an attempt to expand the number of feeling and thinking objects in the world.

Philosophers like Ryle (Ryle, 1949) and Dennett (Dennett, 1978) have gone out of their way to argue that consciousness is an objective phenomenon and must be studied using methods that are as objective as possible. However, they equated objectivity with monism and believed that consciousness has only one nature. We respect and highly value their contribution to the philosophy of consciousness, but we believe that objectivity lies in looking at consciousness as a multi-natural phenomenon. Probably, pluralism is the simplest and most natural thing that can exist in nature.

Artificial Consciousness

When we talk about artificial consciousness, we mean that an object created artificially has some form of consciousness but lacks the same consciousness as an animal or a human. But what does "same" mean here? Obviously, its consciousness will be different in origin, and therefore, it will not be biological. However, apart from this, it must be assumed that the "formula" of consciousness is preserved. But instead of "bio," this artificial consciousness will feature "techno," that is, technology as an artificial analogue of the living. Therefore, artificial consciousness will have a physical-techno-psycho-socio-cultural nature, according to this logic.

The question may arise as to what the difference then will be between the "physical" and "technical" layers. The physical here refers to the active space-time environment, that is, a special dimension in which conscious processes take place. And by technical we mean a neural network, which is implemented not biologically, but artificially in some form.

We argued above that there is no reason to believe that robots and even the most complex language models have feelings. Despite the fact that they know how to create such an illusion with us. Does this mean that artificial consciousness will be a fundamentally different internal experience for the bearer of artificial consciousness – an artificial person? Will the experience of artificial consciousness in the future involve emotions?

The ability to feel is our connection to reality. No artificial creature has this ability yet. Robots only receive information and are not able to evaluate it without our help. Any biological organism, however, can distinguish "not enough" from "okay" or "too much" in some new situation. This is given naturally in feelings. Therefore, a feeling is not only an information exchange but also an experience of reality. It is no coincidence that Norbert Wiener said that the organism is a signal (Wiener, 1961). But an organism is not only a signal, it is also an observer and a signal translator.

The ability to perceive is another connection we have with the world, but it is fundamentally different. Consciousness does not connect us with reality but connects us with the world of meaning. Consciousness is the "sense" of meaning. We see not things in themselves but their meaning for us. For example, we have the idea that a jug is a container with a handle, so when we see a container with a handle, we cannot perceive it in any other way except as a jug, and if we bring our hand to it, then the hand and fingers themselves automatically adjust to grip the handle to pick up the jug. Our feelings and reactions are already tuned to how to interact with the jug. At the same time, the real thing, which only looks like a jug, could be something else. Consciousness brings our influence into reality and helps us perceive this reality in a way that is more convenient for us, in the form of meanings.

Pure sensations almost never reach our consciousness, hiding behind the signs of things. And the signs are already connected with the meanings that we have invested in them. Although the meaning is twofold. For the most part, meaning is of cultural origin. That is, it is the result of certain socio-cultural processes – our collective creativity. In this sense, consciousness is a connection with the meaning that we ourselves give to things. In the second sense, consciousness is our connection with the patterns of the real world in the form in which they contingently appear at a given moment.

Consciousness creates a projection of the meaning of what is happening to us, but it is feelings that tell us what is real and what is not. In a sense, to understand what is happening, we need to be distracted every time from the associative stream of consciousness and, with the help of feelings, determine the coordinates in which we are located. So, for example, in a dream, we are not always able to understand that we are dreaming precisely because we cannot distract ourselves from the stream of altered sleep consciousness and "grasp" the sensations. But at the same time, in a dream, we still feel something. For example, fear and joy or even pain and pleasure. How is it that these feelings in a dream do not bring us back to reality? Unless we wake up, for example, from fear.

It turns out that some secondary feelings can be caused by conscious processes, and they do not connect us with reality. These are echoes, mirages, and dreams, including nightmares. The same secondary feelings can manifest themselves while awake, for example, in the "Rubber Hand Illusion" experiment (Botvinick & Cohen, 1998), when painful sensations are caused by the false conscious belief that the rubber hand is real. A blow to the rubber hand causes almost physical pain, consisting of a complex mix of sensations and premonitions.

The fact that some feelings arise not as a connection with reality but as derivatives of conscious perception, that is, as a sense of meaning, suggests that the fundamental nature of feeling may not be tied to a biological, that is, to a living substrate. Neuroscientists will tell us that secondary feelings arise from the fact that the same neural areas of the brain that are responsible for perception are involved. And they receive a signal not from the outside world but from our "inner world". Therefore, such feelings are weaker, but are formed, as it were, by the same areas of the brain as the primary ones. That is, they will say that their nature is fundamentally no different.

But in this case, it turns out that for the senses, our "inner world" also represents a kind of reality. We can feel a real blow to the hand and a blow to the hand that we are only aware of "inside ourselves."

You can feel by reflecting reality, and you can feel by creating (finishing) some of your own reality. In this sense, feelings and consciousness have a fundamentally different ontological status. Feelings reflect reality – external or internal. Consciousness is the inner reality itself, coloured and reflected by feelings. Some will say that consciousness reflects, to some extent, external reality since, without external objects, our awareness does not exist. But this argument does not change anything since external objects also do not exist on their own, and also reflect, to some extent, some other reality, for example, the reality of the quantum world. At the same time, in the everyday objects that we are conscious of, there is much more of what we culturally attribute to them than of what is in them, regardless of our awareness.

How significant, though, is our difference from AI, for example, from ChatGPT? Let us analyse this issue based on our model of consciousness as a physical-neuro-psycho-socio-cultural reality.

Table. 1. Comparison of Human and AI

	Human	ChatGPT	
Physics: space-time niche	Additional measurements. Not studied	Absent	Does not match

Neuro: neural (information) network	Biological neural network. Human nervous system	Artificial neural network implemented in AI	Functionally identical
Psycho: feelings, personality	Hierarchy of personalities. Memory and idea of one's qualities. Feelings	Absent	Does not match
Socio: internal and external communication	Verbal and non-verbal communication in different roles	Correspondence	Partially matches
Cultural: meaning	Sign systems	Sign systems	Matches

In accordance with the table, the main difference between humans and modern AI is that ChatGPT does not create additional physical reality and, as a result, does not have a personality.

With this approach, it turns out that the most important criterion for the fact that consciousness takes place is that this consciousness takes place somewhere in a certain space-time environment separate from the one in which external reality is located. That is, consciousness here is separated into a separate space (inner world). Therefore, when we say that an artificial person can have the same consciousness as a biological one, we mean first of all that this consciousness, just like the biological one, is allocated to a separate mental space.

But why doesn't ChatGPT create a separate mental space? Why are we so sure of this? And how would it be possible to produce a device that could create its own mental space?

When a person sees a chair, he does not see a real chair but a chair in his mental space – this is a chair in internal reality. But in addition to the mental chair, there is also the real chair itself. Why do we think that the camera broadcasting information to the AI does not do the same thing – that is, why does it not create a special mental space for the AI? Because there are no physical prerequisites for this. Everything that the camera shoots and broadcasts fits into reality, and we watch how the AI processes it. For nature to be forced to allocate additional space for some procedures, it is necessary that these procedures "do not fit" in the existing dimensions.

But what could be an object that lacks space in reality and requires additional space? In fact, an object may not be large but may contain more dimensions or have a special "shape" so as not to fit into our four-dimensional space-time. Such an object is, apparently, the subject. After all, if we formally describe an object (which is a subject) in terms of objects, then we will never have enough dimensions for this since the subject potentially always adds another dimension.

The special property of the subject is that it makes any result the starting point; it turns a function into an argument. Let's try to imagine what this looks like using a simple mathematical notation as an example. This is how a function is usually written:

$$y = f(x)$$

This notation means that each argument value is uniquely associated with one or more function values. For example, "If A, then B." But even if a function has not one but several solutions, it doesn't matter, they are always the same. This corresponds to the properties of an object, such as the ChatGPT language model, and essentially means automaticity. The same reasons lead to the same consequences.

Probably 99.9999% of our abilities, including such exotic things as memory and intuition, obey the same "If A, then B" principle. But some very small percentage of our being – consciousness – somehow manages to make an argument out of a function. Today it could be "If A, then B," and tomorrow "If A, then C," and so on. For the subject, "y" also becomes an argument. And then the function will be something more general, say, "z". And we will write:

$$z = f(y) = f(f(x))$$

But the conscious subject makes a new leap, and "z" also becomes an argument and goes to the right inside the function bracket. And this process continues without restrictions. Consciousness constantly takes as an initial object what has just become a result for it. In fact, this means that the function loops back on itself and becomes like this:

$$f=f(f(f(f(f(f(f(f(f(f(f(f\ldots$$

or simply

$$f=f(f)$$

In other words, a certain formalised space arises (or a combination of spaces, which we call a "non-system"), in which the same cause can lead to different consequences. Objects like $f=f(f)$ or $x=x(x)$ potentially contain more possible relations than reality. They are more complex than reality and, therefore, cannot "fit" into it.

Potentially, a subject has unlimited complexity. However, in practice, "subjectivity" is a level of complexity at least one greater than the complexity of the logical environment in which it is considered. In this sense, the subject is a logical category, and we can judge the subjectivity (of the same ChatGPT) only by the logical complexity of the relations with which it operates.

This brings us to two concepts of complexity. As we know, ChatGPT is capable of handling very large and complex data arrays. This language model has long been ahead of humans in how complex the products of its "inferences" can be. Therefore, we are talking about some other complexity. The two types of complexity can be described as follows. Complexity (1) is the complexity of various relationships within a system, such as in Pic. 3.

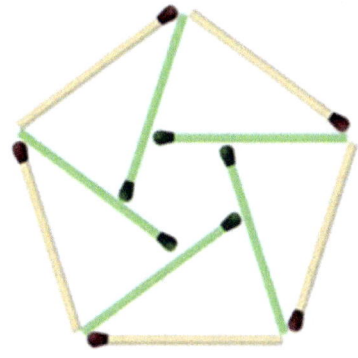

Pic. 3. Complexity (1) – the complexity of various relationships within a predetermined logical environment.

If you set yourself the goal of constructing an object from ten matches, such that it has the largest number of different relations, then you will get some kind of object like Pic. 3. Although the final result will depend on how exactly we count the number of relations. What is fundamental here is precisely that the logical environment in which the object is built does not itself change, and the calculation rules are fixed and unchangeable.

But there is another type of complexity, which lies precisely in the fact that the logical environment itself becomes more complex.

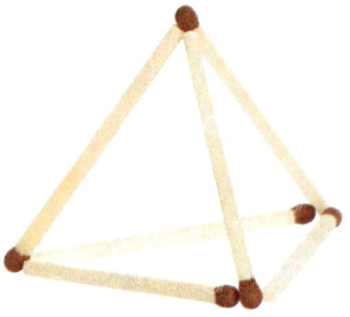

Pic. 4. Complexity (2) – complexity associated with the increasing complexity of the logical environment.

If you set the task of assembling a figure from six matches, in which there will be four identical equilateral triangles, then usually a person first tries to solve it on a plane. But on a plane, this problem has no solution. The solution is found only when a person complicates the logical environment and thinks about a three-dimensional figure. Then he easily guesses that this is the tetrahedron Pic. 4.

So, consciousness probably allows us to artificially complicate the logical environment for considering objects. That is, consciousness is access to complexity (2). But it's not that simple. We conducted a small experiment and asked ChatGPT to solve the match problem. It, just like an ordinary person, began to solve it on a plane and said that there was no solution. But then, after a hint, it solved this problem quite quickly. Again, just like an ordinary person. What's the difference?

The transition from considering a plane to a three-dimensional problem is possible in several ways: because of a hint, as a result of association and memory, and heuristically.

Table. 2. Solving the complexity problem (2) using the match problem as an example.

	Gets a hint	Remembers	Finds an association (guesses it using an example)	Heuristically comes to the idea of three-dimensional space
Child, unfamiliar with the concept of three-dimensional space	Yes	No	Yes	No
Adult	Yes	Yes	Yes	Yes
ChatGPT	Yes	Yes	Yes	No
Artificial Consciousness	Yes	Yes	Yes	Yes

Both the adult and ChatGPT are familiar with the concept of 3D space, which is why we see the same behaviour in them today when solving this problem. They can solve a problem with the help of a hint or an association or simply by remembering the solution. But if we imagine that we are faced with both a person and a language model who are not familiar with this concept, then a difference arises. A human (and our imaginary, artificial human) can heuristically arrive at a solution by intelligently reaching the concept of three-dimensional space, but ChatGPT cannot do this unless it has been trained on the concept.

The transition from a flat model to a three-dimensional one, on the one hand, is a complication of the logical scheme of consideration. On the other hand, simply any arbitrary complication of the model does not lead to solving the problem. After all, the concept of three-dimensional space is somehow and to some extent connected with reality. That is why a person can comprehend it on his own, but ChatGPT cannot. Therefore, it turns out that a biological person's heuristic advantage is not to create just a concept of three-dimensional space but to create a concept that corresponds to some extent to reality. Therefore, consciousness is both a complication of a logical scheme and a complication of it in such a way that it corresponds to reality to some extent.

For consciousness, the logical structures of language that Chomsky described are not enough (Chomsky, 1957). After all, this is only complexity of type (A). To create an artificial person, namely a "device" that would have artificial consciousness, it is necessary to provide him with both skills, that is, (A) the ability to complicate logical environments and (B) the ability to relate them to reality. In this sense, leaving reality should also mean losing consciousness. Not only because consciousness, in this case, ceases to "reflect" the external world, but also because, in this case consciousness itself as a space-time system ceases to exist. Then, we can say that ChatGPT has some form of consciousness, but a consciousness that is not correlated with reality without human assistance, and that is why ChatGPT does not form a mental space-time environment.

Let us discuss, however, how these two skills can be implemented at the "device" level.

(A) Ability to complicate logical environments

Let's formulate some hypotheses. An N-ary logical environment accommodates only N-ary predicates. The (N+1)-ary predicate complicates the N-ary logical environment.

In a single-place logical environment, we can give objects any simple properties that do not depend on other objects. In this environment, identity and negation are possible. In this environment, one-place predicates are possible, for example, "X person" or "X is greater than 5," or "X is not green," but the two-place predicate "X is greater than Y" or "X is not the colour of Y" is impossible. The more variables there are in a statement, the more logically complex it is. What is important is that more complex predicates are not reducible to the sum of simpler ones. The statement "X>Y" cannot be replaced by an arbitrarily large set of statements like "X>5".

Now, let's imagine that any of our concepts is also a predicate with some locality. Depending on what type of relationship is described by this concept. Any non-composite quality is a one-place concept.

The quality of "being green" is a one-place concept. The quality of "having colour" is also a one-place concept. The quality of "having a coordinate" is also a one-place concept. But the compound quality "to have colour and coordinate" is already a two-place concept. For example, a point in a picture has a colour and a coordinate. From this, following our logic, we conclude that with the help of multi-coloured dots in the picture, we can convey relationships only in the form of two-place predicates (concepts). Painting formally conveys objects with simple and compound qualities, that is, objects and the relationships between them.

Here is a running hare, and here is a crawling tortoise, and they compete. That is, "to have the quality of running" or "crawling" is a one-place predicate, and to have the quality of "taking part in competition with each other (with someone)" is already a relation or a two-place predicate.

It is important that any two-place predicate is simplified to a one-place predicate by artificially "convolving" one of the qualities. And, say, the quality of "compete with someone" is simplified to simply "compete," which looks similar, but the first is a two-place concept, and the second is a single-place concept. This means that some concepts may look like simple qualities but be compound.

Thus, we recognise the concept of "run" as one-place and the concept of "compete" as two-place (like predicates). Therefore, in the language, some words look equal but are not, although they form a hierarchy according to locality. For example, the quality of "being victorious" is a tripartite concept, since it implies "competition" and victory, that is, only three sides. The relation "is in competition" is two-place, and the quality relation "being the winner" already contains three places: the competing parties and the winning party. Let's say: A has the quality of "being victorious in the competition between A and B."

Unlike the simple summation of points in an image, where the role of syntax is played by the human imagination, in the text there is a form of formal addition of places. This form is a sentence.

More complex relationships that we capture in painting are thought out at the level of fantasy and imagination but are not formally expressed in painting. However, they can be transmitted using video, since video contains a third gradation of quality – time. A point not only has a coordinate and colour but also a mode of appearance in time on the screen. Therefore, we can formally observe that the competition between the hare and the tortoise initially proceeds in favour of the hare, but then the tortoise still wins without resorting to fantasy. Video, although artificial, showing individual dots of different colours, formally conveys a certain "life" of processes or relationships.

That is why video conveys a realistic life and involves us in the viewing process. In addition to complex relations, a three-place predicate can also form a relation of subjectivity within the predicate itself. This forms higher-level metalanguage concepts. "I won," says the tortoise. But we are not talking about the one-place predicate "there is a winner," but about the three-place predicate "the one who is now saying these words is the winner ". That is, direct speech is a minimally three-place predicate since it already contains text and context. Moreover, the text itself with its relations is, in this example, a one-place predicate (there is a winner), and the measurement of the context that this is direct speech (the one who is now speaking these words) makes it three-place.

If direct speech forms even more complex relationships within itself, then this makes it possible to further increase the locality of predicates, creating layers of texts within texts. For example, my sister's phrase: "Mom, he doesn't listen to me!" in relation to her brother, is an even more complex construction. By breaking it down into predicates, you can get the following. The very content statement here is the threefold quality: "to be disobedient." It is tripartite because it implies the following: he "is the one whom I encouraged to do such and such an action but did not do it." The context of direct speech here is also more complex: I am "the one who is saying this now to her mother. I am reporting that someone (my brother) is...".

Now, we come to the key point. Unlike AI, human computing resources are limited. Using the examples above, we showed that the subject carries out the logical complication of texts but at the same time uses methods of simplification or "convolution" of concepts. These methods include:

1. Giving new names to more complex objects that are part of or composed of simpler objects that already have names;
2. Giving new names to more complex relations or qualities that are part of or consisting of simpler relations or qualities that already have names;
3. Convolution of a set of objects and relations due to direct speech (that is, due to the introduction of speech subjects);
4. Etc.

Due to "convolution", the subject gets the opportunity, with the help of predicates of the same area, to operate within the framework of more complex logical spaces. That is, an increase in the complexity of the logical environment is organically carried out through the complication of concepts. However, the complication of concepts through their "collapse" creates "collapsed" places within the concepts, and this has at least two consequences.

Firstly, there is uncertainty about these concepts. "Collapsed" places in a concept require interpretation by subjects since their content is not filled in the concept itself. Let's say the quality "to be naughty" contains at least one "collapsed" place. If someone says that "Peter is disobedient," then the place that contains information about whom Peter disobeyed is not revealed. The subject who is told that "Peter is disobedient" interprets it in such a way that Peter disobeyed the one who said it. But this is an assumption for which the subject needs to "complete" the meaning of what is being said.

Secondly, due to the growing complexity and uncertainty of concepts, the complexity of the process of correlating statements based on them with reality also increases. This is not only due to the complexity of interpreting what is being said but also to the difficulty of determining evaluation metrics about what is being said in practice. Let's say how should one treat the statement that "Peter is disobedient"? To what extent should you trust him, and how can this information affect Peter's future behaviour? How can we understand from Peter's behaviour that this predicate is true to some extent in reality? The more complex the sign system of communication becomes, the more the subject is forced to "complete" the communication himself. People don't notice this in everyday language, but literally, every word and every sentence can be understood in at least two or three ways. As a result, consciousness is necessary to, if necessary, "expand" predicates and supplement each received message with numerous pieces of information, without which the messages are simply incompatible with reality.

(B) Ability to relate logical diagrams to reality

Correlation with reality consists in answering the question: "to what extent is this possible in reality?" Very often, we are not aware of how important it is for us that something is realistic. For example, feelings such as envy or sympathy are highly dependent on the assessment of realism. If we see that someone has achieved something that we quite realistically could have achieved but did not achieve, this causes envy. But when we observe people receiving something that we could never have received without a conventional miracle, it is unlikely to have a strong emotional impact on us.

Also, with sympathy, we are unlikely to feel great sympathy to the degree of attachment for a person who, in our opinion, could not be imbued with reciprocal sympathy. A person always seeks mutual understanding with another, but if it does not seem realistic to us with some person, then we will not strive for mutual understanding. On the contrary, the feeling that mutual sympathy or understanding is realistic significantly fuels interest. Or maybe this is generally a key factor moving us towards reducing the distance in communication, and our sympathy is simply faith in the success of communication.

A person constantly evaluates the realism of something, distinguishing "possible in principle" from "really possible." And this assessment makes everyone who they are. For example, people who believe that winning the lottery is possible will often buy lottery tickets.

To impose a logical framework on reality, however, it is necessary to have a model of reality. Therefore, by calling consciousness a sense of meaning, we mean that consciousness is the ability to relate a simulated internal reality to the interpretation of the received sensory experience. That is, "meaning" and "model", in this case, are essentially the same thing. The realism of something is also assessed by consciousness in accordance with the model of reality.

Let's imagine an AI that we have taught to understand the world, constantly complicating its concepts. Such knowledge, however, will be "dark" if its results are not correlated with practical experience. AI can form an extensive system of concepts; however, it will not only reflect reality but, more importantly, model it.

The human body draws its first models of reality from instincts. The older a person gets, the more complex his motivation and methods of implementing models become. Conscious checking of connection with reality is fundamentally different from intuitive (unconscious) testing. Intuition or instinct may suggest that buying a lottery ticket means a real chance of winning, but conscious testing may override this pattern. So, how does one check the connection with reality on a conscious level?

Intuition may tell a person that this person is not worth dealing with, but consciously, he can convince himself that his intuition is wrong and ultimately develop trust. For us, it is not important whether the conscious decision is correct or not. It is just an assessment of the effectiveness of the model. It is important to understand how consciousness relates to the model. Is there a choice between various already existing intuitive models, from which consciousness simply chooses the most suitable one, or is consciousness capable of building its own models?

We will proceed from the fact that the conscious model of the world is a completely special type of model. This is not a huge archive of data and information that is at the disposal of the subconscious, but a very limited "picture" of what is currently recognised as possible. We can say that the conscious model of the world is pre-consciousness, that is, everything that is now ready for immediate awareness or is already being realised.

Collectively, this is a "picture" of models-meanings that consciousness can consciously manipulate.

To be in the preconscious for information means to be at a certain high energy level. A person can consciously direct his attention, but the content of this information is drawn from memory and external sources. Conscious manipulation of models-meanings is what we carry out as an activity of consciousness constantly. The initial "picture" of the conscious model itself arises without the participation of consciousness, but by controlling voluntary attention and mental manipulation of objects (completion), consciousness complements and changes the original "picture".

The key problem here is to understand how conscious manipulations are carried out in terms of the physical-neuro-psycho-socio-cultural nature of consciousness. Where is the source of these manipulations? How does the "movement" of thought begin? What ontological status do these manipulations have?

The peculiarity of this approach is that since the nature of consciousness is considered here as heterogeneous (physical-neuro-psycho-socio-cultural), then its core will be heterogeneous. This means that conscious "movement" can take its source, in accordance with this approach, in any of the beginnings. Let us give examples of each of these sources of mental "movement".

The sources of the conscious process can be:

1. Topological effects of mental space-time
2. Internal neural processes when a certain energy level is exceeded
3. Feelings caused by external events or qualia
4. Communication
5. The meaning of what is happening

This, of course, requires some explanation. The point is that these sources are interrelated, and a comment may follow that, say, communication (4) begins with feeling (3). After all, speech is sound, and sound is a feeling, and if someone tells us something, then before we understand the words and speech, we hear the sounds of speech. Therefore, it is possible to object that sound here is the primary source rather than speech itself.

But this will be a mistake since you can listen without hearing. If we hear something as a sound, this does not mean that this sound (this feeling) is conscious of us. That is, sound can certainly lead to our "awakening" from the "slumber of the flow of associations," but for this, the sound must meet the criteria that lead to awareness. The same goes for communication. Not all communication is carried out consciously. You can speak and answer "automatically."

When we talk about the sources of "consciousness," we mean that each of these sources can trigger a conscious process if there are sufficient conditions for this (second column in the table). Once launched from one source, this process will involve all other specified sources (third column in the table). It would be more accurate to say that this process will become "conscious" only when all environments participate in it.

Table 3. Different sources of a conscious process.

Source of conscious process	Sufficient conditions to start the process	Action of this source when a process starts from another source
Mental space-time	Significant topological effects	Dynamic change in the topology of mental space-time
Neural network	Exceeding the energy threshold for neural processes	Activation of neural processes in different parts of the cerebral cortex

Psyche	Exceeding the threshold of perception	**Active attention**
Communication environment	Meaningful communication that requires awareness	Internal (communication of social roles of one person) or external communication
Symbolic space	Meaning that evokes a conscious response	**Performing signed transformations and calculations**

The comparison of a logical scheme with reality in this approach thus means the synchronicity of "active attention" and "sign transformations and calculations." It is also important that sign transformations here are synchronous with the dynamic processes of mental space-time. This means that the cultural environment of consciousness is in relationship with the physical environment of consciousness. Semantic transformations influence the physics of conscious processes and vice versa. It is this integration of five environments of different natures that makes consciousness still the most difficult object to study.

Preservation of Personality in an Artificial Body

How to transfer personality from a biological body to an artificial one? There is already a lot of discussion about this. The main problem here seems to be how to distinguish the original personality from its copy. For example, if you create a neural network that exactly replicates the brain of a real person and then connects it to an artificial body, you will actually get an artificial double of the person and not the person himself in an artificial body. The original personality will still exist. But the task is not to create duplicates of people, but to try to make people themselves eternal in new artificial bodies.

In fact, if we take into account one of the main properties of consciousness, the problem of double personality fades into the background. This property of consciousness is its ability to occupy the position of a supersystem in relation to the systems in which it exists. Simply put, consciousness always strives to take the position of "boss" in order to control the entire system, no matter how gigantic it becomes. In essence, our brain is a very large hyper-system consisting of a large number of systems. It is only thanks to consciousness that our brain "pretends" to work as a single whole, and even then, not always.

We have previously described this property of consciousness in more detail using various examples. Here, we will not dwell on this in detail. We will only note that this "will to control and power" that consciousness exhibits can be used to preserve personality in an artificial medium. If we were to artificially expand our brain, adding, for example, a super-cortex that would surpass our cerebral cortex in complexity and volume, then our consciousness or our personality would gradually establish its control over this new neural territory. This, for example, happened throughout the biological evolution of the nervous system. The animal brain became more complex due to the emergence and development of new areas, and the "control centre" of the animal moved to more and more new areas.

If the new artificial area of the brain turns out to be more powerful in the information sense, then consciousness will move the "control centre" of the personality to this area. In the future, after some time, the older biological areas of the brain will lose priority in the formation of personality. Therefore, to preserve personality on an artificial medium, it is only necessary to create a physically more complex brain than our own. Our personality itself will gradually change its "registration" to a new one.

Three Technologies

All the technologies discussed below can improve the human body directly or indirectly. In fact, all our technologies are dependent on the human body, namely on the characteristics of our sensory organs and actors, on the materials of our body, and on our needs.

(1) Neural Interface

Let's take, for example, the space program, which began in the middle of the last century. It was obviously necessary to consider that astronauts are people. This means that they need air, water, food, a special temperature regime, a special pressure regime, humidity, and so on. Therefore, the space program was divided into two subprograms – one with the participation of astronauts and the second with an autopilot. What are the comparative results of both routines? Autonomous vehicles have reached at least the boundaries of our solar system, while human-piloted spacecraft have not advanced further than the Moon. The difference in the scale of subroutines is approximately ten thousand times.

As you can see, the need to take into account human characteristics greatly slows down the space program when it requires the participation of astronauts. Now let's imagine for a moment that scientists have developed a technology that allows us to change the human body and make it less fragile, more resilient, and less dependent on environmental conditions, air, water and food. This would speed up man's conquest of space enormously and would make it possible to visit other planets of the solar system.

Protagoras said that "man is the measure of all things." Today, we would say that the human body is the measure of all our technologies. Ultimately, we adapt all our tools and devices to the characteristics of the human body. Therefore, even small changes to our bodies would lead to a very serious revision of what our tools should be in the broadest sense of the word.

The first technology that can change a person is called a "neural interface". Today, this is still a niche area, but it is likely to become the main driver of the economy in the next century. The essence of this technology is literally to create devices that a person can control with the "power" of thought. It sounds unscientific, but if you think about it, this is exactly how we control our bodies – with the power of thought. And there is nothing unusual about this. From a material point of view, thought is the electrochemical interaction of neurons. This is probably not all that thought is. However, this is quite enough to create a neural interface.

"Reading a person's mind" means doing two things: receiving the desired neural signal and deciphering it. First, it is necessary to receive a signal directly from those neurons of the human neural system that are excited in connection with the thought that we want to recognise. For example, a person with an amputated arm can send neural commands to move the lost limb. It will not succeed, but the corresponding signal will be generated, and it can not only be detected but also amplified.

To capture the signal, three options are used: non-invasive, minimally invasive and invasive. In the first case, we are talking about the well-known technology of electroencephalography, when signals are read from the outside without penetrating the nervous system. The other two options differ in the degree of sensor penetration. The invasive method of penetrating directly into the brain is the most accurate and allows for much better differentiation of signals, which is important for the next stage – decoding thoughts.

So, the second thing to do is to decrypt the received signal. This means, for example, understanding exactly how the already mentioned imaginary person is trying to move his hand: whether he wants to bend only his thumb or, say, clench his entire fist. The signal for each type of movement will be different, and there is some language of correspondence between neural signals and the motor commands they give. This language is unique to each person and cannot be deciphered on the basis of any general principles, but, fortunately, it is often quite simple to record it by asking the person to send certain signals and establish correspondence.

To some extent, this is the same as taking fingerprints and then recognising their owner by the prints. The same goes for every single "thought" of a given person. To decipher it in the future, it is necessary to record its "neural fingerprint."

The good news is that you can't yet read a person's mind without doing some research with them first. However, this does not always mean that a person must cooperate. For example, you can show the subject different images and read what characteristic signals arise in different areas of the visual cortex. In the future, this would make it possible to record whether the subject thinks about these images. In this case, cooperation on his part would not be required.

Initially, the neural interface was developed for medical purposes. Using this technology, partially and completely paralysed people can control a variety of devices, from a computer mouse and a power wheelchair to a bionic prosthesis. Unfortunately, brain-computer technology is at the very beginning of its journey, and until recently, it was not possible to help even the outstanding physicist Stephen Hawking.

Stephen was almost completely paralysed, and his brilliant brain was imprisoned in a cage of his own body, unable to share his thoughts. He was brimming with the most sophisticated ideas of modern physics, but he could neither write with a pen nor type on a keyboard nor speak. Various devices have subsequently been developed specifically to allow him, albeit very slowly, to work with text.

Initially, Hawking retained some motor activity in his finger, and he could control a special device on the armrest of his wheelchair. But over time, even this small motor activity was also paralysed. However, his cheek still retained minimal activity, and he could make movements at will. So, a sensor was developed especially for Hawking that reads information from the earpiece of his glasses. With just one facial muscle, he gained access to the whole world. He could search the Internet, write text, and even "talk." A special voice synthesiser spoke for him, which voiced the text previously printed by the physicist using his cheek.

But the mobility of the cheek also constantly deteriorated, slowing down the process. And although the scientist got the opportunity to write and publish a book in this way, the joy, unfortunately, was short-lived. The mobility of the cheek also deteriorated sharply. Hawking was fitted with an electroencephalograph to try to read his mental commands directly from his brain. But the signal turned out to be too weak, and this opportunity was not realised. However, the decade that separates us from the time of these events has not been in vain for neural interface technology. There is reason to believe that today Hawking could be helped much more effectively.

To understand all the revolutionary nature hidden behind the modest name "neural interface", it is necessary to reflect on what a material personality (hereinafter simply "personality") is and how it relates to the body. As the outstanding American philosopher of consciousness Daniel Dennett said, a person can be understood very narrowly and very broadly, and you never know which scale to choose. And indeed, why, for example, is a hand a part of a person's personality, but a mobile phone is not? What can we say about the prosthesis?

If we consider a personality very broadly, then its halo may include constantly used personal belongings, important tools and even a home, as well as key social relationships, outstanding professional achievements, and so on. In some very broad sense, during the life of Peter the Great both the Russian fleet and the city of St. Petersburg were part of his outstanding personality. They were inextricably linked with him. However, in this sense, the personality turns out to be maximally deindividualised. A person here acquires the features of an entire institution and turns out to be almost completely determined by external factors.

If, on the contrary, we consider the personality very narrowly, then it turns out that our own body is not completely included in it. If personality is consciousness, an individual substance that freely acts at its own discretion, then it should be sought exclusively in the human nervous system. The hands only receive a signal from the brain, so the hands themselves are not part of a person in this sense. Therefore, personality is a special, narrowly specific neural activity of the cerebral cortex and not our entire body.

What, however, is the significant difference between both approaches? Where is the criterion? The main criterion we have used here is not so much the scale of consideration as the way the individual controls his instruments. In the case of a broad approach, we believe that any management, if it is very strong and systematic, is suitable. Therefore, things, relationships, and sometimes even entire cities and institutions entered the aura of personality. In the case of a narrow approach, on the contrary, we decided to consider a personality to be only its very core, which is not controlled by anything but controls itself; that is, the controlling entity in its pure form. Therefore, even our body did not become a part of such a personality.

But a third approach is also possible, in which a personality is everything that is controlled by the thought of this personality.

In this case, it turns out that the personality will consist of a human body and a neural interface that complements the body. In this sense, it turns out that a bionic prosthesis, controlled through a brain-computer interface, is part of a person's material personality. And this is exactly how it will be perceived by the person himself. But this applies not only to bionic prostheses but also to any devices that are constantly controlled through a neural interface, be it a motorised wheelchair, a computer mouse, or even a home theatre or car.

A person will feel that these devices are the organs of his "larger" body. Moreover, he will attribute the abilities of these devices to himself, considering the skills implemented by external processors to be his subconscious. Having learned to ride a bicycle or type on a keyboard with ten fingers, we no longer wonder how this happens. The same will happen with neural interface devices, simply bypassing the training stage.

The neural interface in the future will expand and radically change our understanding of what the body and personality are. And what seems wild and fantastic to us today will become a reality very soon since this technology is developing very quickly. The usual body will be replaced by a "big body" that runs off various devices. It turns out that people's "larger bodies" will be able to intersect with each other if two or more people use the same device together.

(2) Machine Learning

The human brain is extremely plastic and is capable of learning to work with a wide variety of tools. Therefore, the growth potential of the "big body" is very large but not unlimited. What makes it truly limitless is the second technology – machine learning. This technology consists of "artificial skills" or skill automation. It allows you to increase a person's capabilities without using the resources of his own brain. If the neural interface expands the human "body," then machine learning expands the human "neural network" that controls this body.

For example, for the previously mentioned Stephen Hawking, a system was developed that sped up the typing process. The fact is that the scientist could only make one simple press, and therefore, he had to stop the cursor, which automatically moved through the alphabet. To type a word of five letters, you had to wait until the cursor went through the entire alphabet five times to the desired letter. To speed up the process, the developers introduced a bit of "machine intelligence" into the system, which suggested to Hawking the words he frequently used by their first letters. Today, this ability of a predictive text editor is well known to us all.

We are all already quite actively using this technology today, which is usually called artificial intelligence, although it is more accurately called machine learning. There is nothing mystical in machine learning, nor in the neural interface. This technology works literally on the same principle by which the neural networks of animals and humans have always been trained. And we have implemented this principle at the machine level, although so far less efficiently. It's called backpropagation. Its essence lies in the fact that, during training, the initial coefficients of the neural network are slowly adjusted. After a very large number of training sessions, the coefficients are adjusted in such a way that the neural network itself begins to do what it was taught. This way, you can automate everything that a person can do automatically. For example, recognising faces, driving a car, writing text, drawing pictures, etc.

In the last century, we were greatly disappointed with the capabilities of artificial intelligence. Unlike nuclear physics, where progress was ahead of science fiction, with artificial intelligence, it was the other way around. There was so much talk about it, but there were so few successes that it seemed that in our hearts, we had written AI into approximately the same section where space exploration had become located or the search for alien civilisations. That is, we included AI in the list of perhaps promising but still very distant prospects.

After the "winter of artificial intelligence", which lasted for several decades, "spring" has arrived. And today, unexpectedly for many, this technology is showing rapid growth. The "skills" of various AI systems based on the principle of machine learning have moved over the past two to three years from the zone of "much worse than a human" to the zone of "almost human-like", "human-level", and even "better than human". We are talking about various abilities associated with recognising and generating content, that is, those areas that have traditionally been a person's strength.

Unlike humans, machines learn more slowly, but progress in learning is always maintained. At the same time, they do not get sick or die and are able to accumulate "skills" around the clock and tirelessly. If they have learned something, then these skills are no longer lost and can be easily replicated and distributed between machines. After all, for a machine, a "skill" is just a database or code that is transmitted via Wi-Fi.

If this technology continues to be developed, then in a fairly short time, AI will be able to do so much, so quickly, and so well that it will fundamentally change our economy. But most importantly, this technology will help change the person himself, expanding our ability to control our "big body." The human brain has limitations in the number of tasks that it solves simultaneously and efficiently, has memory limitations, rather weak computing abilities, etc. But science can help here, too.

Neuroprosthetics technology, which to some extent combines a neural interface and artificial intelligence, looks promising. A neuroprosthesis is an electronic implant that allows you to restore some function lost by a person, for example, in the simplest case, hearing. Let's say a cochlear implant consists of a microphone that picks up sounds from outside, a processor that processes this signal, and an implanted unit that stimulates the auditory nerve fibres in the cochlea. In this way, the function performed by the eardrum is restored.

This simplest example of a neuroprosthesis, however, clearly demonstrates the structure and potential of using neuroprostheses in general. This potential is determined by several factors, and the most important is probably the ability to integrate electronic processors into the structure of the human nervous system. This would make it possible to build up not only the "large body" of a person but also to expand our intellectual abilities inherent in nature by integrating additional computational modules into the nervous system. In a cochlear implant, such a module is minimal, but the fact that this is possible in principle opens up enormous prospects. Today, we are already seeing large companies engaged in such developments, and tomorrow, the neuroprosthetics market may take the lead.

Intellectual ability, however, has its limits. Just as we cannot endlessly increase "peripheral devices", since this requires increasing the power of the control processor, so the increase in design capacity is somewhat limited. And this is a management problem that we don't understand very well yet. Let's try to figure it out.

Brain-body interaction is, in many ways, a management process like any other. And for management it is important to maintain certain structural and functional proportions. For example, the power of a processor that controls a washing machine is unworthy to control the autopilot of an aeroplane. A tired driver is likely to drive more slowly than an alert driver, as it will be more difficult for him to process information about the situation on the road at higher driving speeds. That is, to manage a large number of independent processes, appropriate design power is needed.

However, the endless increase in design capacity is also problematic. Why? Because there may be a conflict between the functions of different control processors. Our brain, among other things, is an integral organ, which, however, consists of many independently operating control centres. If these centres conflict with each other, we experience discomfort, and in extreme situations, even a split personality may occur. The main arbiter in this confrontation is a special level of mental activity, which we call consciousness. We still know little about consciousness, but it is already clear that this is a special activity that cannot be reduced to calculations.

The fact is that independently operating centres of our brain can "agree" with each other themselves, only in a language of numbers that they understand. In other words, if the calculations show some obvious picture, such as: "Now we need to get off the couch and go to the grocery store. Otherwise there will be nothing to eat," then the decision can be made unconsciously, and the person will not have to resort to a conscious analysis of the situation. A lot of things in our lives happen like in the Libet test (Libet, 1978). But a situation like: "I need to stop at one glass of wine and go home" or "it's time to wake up and go to work" can be more difficult and cause resistance.

Our consciousness, like an orchestra conductor, is able to give special access to the control of our body to individual control centres, depending on the situation and our will. This is a special skill that no robot has yet and which evolution has been creating for billions of years. This is why, for example, the ability to behave consciously is not always associated with a person's intellectual abilities. In particular, it is known that people who are deeply involved in mathematics can sometimes be absent-minded in everyday life, not paying due attention to everyday problems. The fact is that consciousness, on the one hand, and the ability to make calculations, on the other, are fundamentally different things. Consciousness is responsible for prioritising (allocating attention) in new and unknown situations rather than making calculations to solve each individual problem. The more diverse calculation modules a system contains, the more developed the consciousness function must be to control this system.

The most difficult thing about consciousness is that it is a cognitive function. Among other things, this also means that this function is only required when first encountering a new situation and when learning. Our nervous system strives to automate all actions, reducing energy consumption (consciousness spends a lot of energy). Therefore, when our natural neural network learns something, we begin to do it "automatically," sometimes without even realising that we are doing it. The situation is the same with artificial neural networks. They only need our attention during the learning process, and then they do everything themselves automatically. But if some new situation arises, the human consciousness is again required to retrain and solve the problem.

Let's imagine such an unlikely but possible situation. You drive to work along the usual route, the same as every other day. But this time, an unusual situation occurs in the centre of the city, on a busy street. A panther comes out onto the road right in front of you. And you see that the panther is moving towards the people who are standing on the other side of the road. You realise that you could run down the panther, thus preventing it from harming a group of people. But at the same time, you don't know anything about this animal: Is it really going to attack the people, or is it a harmless, tame animal that is scared itself? Any extreme, but still possible, situation can be imagined demonstrating the fact that no AI system operating a machine in such a situation would be able to act intelligently unless it had been trained to do so first. A person here also does not have many chances to make the right choice, but he certainly has one.

We will discuss the pros and cons in detail below, but here, we consider it important to mention the third technology, which will radically change our lives in the future. And perhaps it is this technology that will give us immortality.

(3) Artificial Consciousness

Scientists sometimes mistakenly confuse this phenomenon and call it strong artificial intelligence, but these are fundamentally different things. By strong artificial intelligence or general intelligence, we mean an AI that can solve any problem at all but unconsciously. By artificial *consciousness,* we mean a conscious AI that is capable of solving any problem at all consciously.

The difference, at first glance, is speculative. In both cases, we are talking about a certain environment that has the potential to solve any problem without being tied to specialisation, without restrictions on the area, etc. It seems that only the method of achieving this goal differs. In the case of strong artificial intelligence, this is a "soulless" program, like ChatGPT, and in the second case, it is a conscious agent aware of the world, including feelings and experiences. The issue of whether robots feel is dealt with below, but now we will discuss the question of why the technology of artificial consciousness is fundamentally important and what is valuable about it.

Already today, based on machine learning technology (the second technology on our list), it is possible to train a neural network to answer questions in the same way as a specific person does. After the death of this person, it is theoretically possible to continue communicating with this neural network as if the person continues some digital life. At least the neural network will create such an illusion. And this illusion will be stronger, the better the neural network has been trained. However, such a digital "cast" of a person's soul will be like a reflection in a mirror. It will only reflect a once-living person, albeit in a very complex form.

You can imagine it like this. A portrait or photograph conveys a flat projection of a person's appearance. The sculpture reflects the shape of his body. From a copy of a person's body, made with great accuracy, it would be possible to determine, for example, how a person would move in each situation. All sorts of physical tests could also be performed on such a copy. A neural network that answers questions like a person is something like a semantic copy of a person: a cast of his way of thinking, made at a moment in time. By asking this neural network questions, we would also conduct tests with it – not physical, but semantic. But this does not change the essence. A neural network trained to respond like a specific person is just a "portrait of a soul" and not that soul itself.

If strong artificial intelligence technology is developed, it may become possible to create an electronic copy of a person – one that not only reflects the semantic structure of the person's soul and answers questions as they would, but also acts as they likely would in unfamiliar situations. Such an intellectual agent could, for example, complete a book left unfinished during the author's lifetime, or respond to original, authentic questions the person never considered and had no formed opinion on.

Such an agent would be capable of conducting outwardly intelligent activities and even developing. However, it would not experience any emotions and would not have an inner world, that is, a soul. It would no longer be a thing but also not yet a living being. The ontological status of such a being could be characterised as a self-propelled artificial intelligent agent. Philosophers are still debating whether such an agent would be capable of behaving completely human-like in appearance or not. There is, in our opinion, an erroneous opinion that human abilities do not depend on consciousness (the presence of internal vision) and that consciousness, in accordance with this opinion, is only a random accompanying optional effect. We will discuss this below, noting here only that this "optional effect" is extremely important to us. It allows us to experience and feel in the broad sense of the word. "Why does the universe need my worries and torment?" – the person asks, but this is the main thing we have.

In the case of the development of the technology of artificial consciousness, one could already say that the artificial agent acquires the status of an animate object, a personality – an artificial person. This would open the way to using such technology to copy a personality or transfer the personality itself to an artificial medium. And this, in turn, means actual immortality.

It is important to note that copying or duplicating a personality today looks to many researchers as a more accessible technology of the future than transferring a personality to an artificial medium. But this is an unsubstantiated opinion, based only on our current rather meagre ideas about the nature of personality. Indeed, if we proceed from the machine learning technologies available today, this will explain such a point of view because machine learning is essentially a technology for copying skills.

It seems that the development of artificial consciousness technology would radically change our ideas about what a person is. And this technology would certainly give us the opportunity to endlessly explore the world around us in new immortal artificial bodies. Although other problems would arise that we do not yet even suspect, we can try to imagine them.

So, three technologies – neural interface, machine learning and artificial consciousness – will completely change people's lives soon. They will make a person immortal, but in return, they will deprive him of his naturalness (natural origin).

Biological Man: Advantages and Disadvantages

If we consider a biological person as a device and compare it with the devices that we can produce, then a biological person has many advantages. These advantages, however, should be divided into three groups: (1) The advantages that we previously had over technology but have lost; (2) Advantages that we still have today but most likely will soon lose; (3) Advantages that we may never lose, or at least not soon.

Lost Advantages

Here, it is proposed to look broadly at all our properties and abilities, considering them as biological, natural, and evolutionary tools in solving problems. And then it is proposed to compare them with artificial instruments.

So, let's put it this way: our artificial tools have become better than our natural tools for five reasons:

- Due to the use of the functional properties of materials;
- Due to the application of designs and structural advantages of the devices;

- Through the use and transformation of different types of energy;
- Due to the ability to program a sequence of actions;
- Due to the ability to train devices.

So, we lost the advantage where our artificial tools turned out to be sharper, stronger, faster, more durable, and so on because the tools used more functional materials than the materials that make up our bodies. The stone knife turned out to be stronger than our teeth and nails, and the club or stone axe was stronger, heavier and tougher than the hand. Wood turned out to be better for moving on water, and animal skin was more reliable for retaining heat. Copper, iron, steel, glass, plastic, composite and nano-materials and so on, gradually replaced the materials of our body in various processes.

But materials aren't everything. Even today we use our body as a tool, simply protecting it from external influences with the help of various protections, such as wearing gloves or glasses. That is, we have artificially improved the materials of our "cultural" body. At the same time, our body has long been an ideal design for various tasks. Evolution taught us to walk, run, swim, climb trees and mountain slopes, and also use our hands as a complex system of blocks and levers. Until now, the human hand is an unsurpassed manipulator when it comes to, say, plumbing work or, for example, rock climbing.

The hand is a universal manipulator, and it is still indispensable when some tasks consist of diverse actions. But in relation to narrow tasks, man has created many more effective structures, manipulators and levers. By replacing joints with wheels, bones with beams, and muscles with cables, we got very strong artificial "arms" capable of breaking rocks, digging trenches, moving large loads, and so on.

For a long time, we used the energy of our body for all these purposes. We have exploited our evolutionary advantage in efficient energy distribution. In particular, upright walking made it possible to spend less energy on moving, and resources were freed up to use arm strength. Structures have become amplifiers of our strength, thanks to levers and pulleys. But over time, this became not enough for humans, and they began to use the energy of water and air, the power of animals, and then the power of steam, the energy of explosions, electricity, nuclear reactions, and so on. This turned out to be much more effective.

So, devices have basically got better than us due to more efficient materials, more efficient designs, and more efficient energy sources. For example, a modern excavator can replace at least a hundred people with shovels and picks in terms of speed and volume of tasks performed. But at the same time, the excavator still needs to be controlled, that is, by an operator. This is because the work of an excavator consists of diverse actions. And for a long time, we had no equal in controlling the action programs of instruments. However, we invented cybernetics and devices capable of storing sequences of commands and then implementing them at the right time.

So, we lost our advantage in managing processes that we understand well. Let's say we understand very well how to boil water, so the kettle turns off on its own after reaching a boil. The teapot processor implements a very simple program. Although, in essence, any program in the field that we understand very well can be implemented according to this principle. We created an autopilot for an aeroplane and a robotic line in a car assembly plant. Satellites in Earth's orbit are automatically controlled. All these processes are decomposed into simpler ones, which we know very well.

And yet, even with sophisticated computers, we have long maintained an advantage where we did not understand the processes very well and could not translate them into programs and a set of instructions. For example, we were better than computers at pattern recognition for quite a long time. Therefore, a lot of time passed from the invention of the autopilot of an aeroplane, where this skill is not required, to the invention of the autopilot of a car in a city filled with pedestrians and other vehicles.

For many years, pattern recognition seemed to us a daunting task, and we could only entrust it to a person. This is because pattern recognition is not broken down into simple commands that we can understand and program into a computer. The same type of task includes creating images based on the task at hand, as well as working with meaning in general.

Alan Turing posed the problem of whether a computer could imitate a human by answering questions as intelligently. This problem, as it turns out, is on the same level. In general, we seem to understand that the computer should respond, but the number of language situations is so large that it is not possible to construct an algorithm for its responses in any foreseeable time. One would simply have to list all possible questions in all possible contexts and provide answers for them.

We were the best at this, too, for a long time. However, the creation of machine learning systems made it possible to solve this problem, and the computer was able to replace us in this, too. Today, AI very quickly, with a high probability of success and tirelessly, recognises images, drives a car, responds to requests from users of an internet service, conducts a conversation, generates complex texts upon request, draws pictures and even comes up with videos. And although these are just the first steps for AI, the results are already very impressive.

As is always the case with technology, if something can be automated in even a very mediocre way at first, over time, the quality of automation increases exponentially. We will discuss how to use AI for a new fundamental breakthrough in technology below, but here it is important to note the following: AI based on machine learning technologies existing today will eventually be able to do better than us in everything that we can do automatically, that is, unconsciously. And this is a huge layer of skills since, sooner or later, we learn to do almost everything in life without thinking.

Advantages that We Have Not Yet Lost

This category contains those strengths of a biological person that can be realised artificially, but this has not yet happened.

In the previous section, we discussed materials and designs. The most interesting thing, as always, is at the junction of two areas. The tissues of our body can, to some extent, be called self-healing materials, although, in essence, they are very complex biological "constructions". Skin regeneration is one of the heights of nature's evolution that has not yet been achieved at the artificial level. We already know how to create very complex materials-structures with unique properties, but they have not yet reached such perfection as ordinary human skin.

Another amazing property of cells and organisms is self-replication, which manifests itself at different levels. The basis of this process is the self-replication of a gene that carries genetic information and acts as a biological "memory" of the organism. In this sense, the regeneration of skin cells is also one of the manifestations of self-replication. But in addition to cells, we can also talk in general about the self-replication of organisms, which is called reproduction.

The great advantage of biological "constructs" – that is, organisms, is that they have, as it were, two levels of processors. One of these processors, in humans, is the brain, and it centrally controls the body in the same way as the processor of any computer through nerve channels. Violation of such a channel (say, a spinal injury) also leads to disruption of the controllability of some organs. But there is another level of processors – cellular. The program of action of each cell is carried out at the level of its gene. Therefore, virtually every cell of the human body has, albeit very limited, intelligence. It is our cells that are responsible for the amazing properties of our fabrics, which have not yet been obtained in artificial materials.

As for reproducing, our technology cannot yet do this either. Our refrigerators do not "give birth" to children. We do not yet know how to create structures that would grow from an "embryo", consuming energy from the surrounding world. But such "devices" would be very useful, say, for sending to Mars or other planets. Transporting every kilogram of weight to Mars is very expensive, and if we could transport there the "embryos" of Mars rovers that grow in situ, this would greatly simplify the implementation of Mars programs.

Advantages that We Will Never Lose

Unfortunately, there isn't much to show you in this section. It is important to understand that either a person will make the development of technology part of their own development, or at some point, technology will begin the path of its own evolution, regardless of humanity.

The fact is that the strengths of a biological person are, in fact, their weaknesses. We love children, but being a child means being helpless and in need of care. We remember our school years fondly, but being a schoolchild means you need basic knowledge and depend on the educational system. Everything that we do is forced. We try to do it with pleasure. For example, we eat with pleasure, although it is just a "refuelling", without which we will die.

The problem is that the most pleasant memories in life are usually the memories of how we cherished our weaknesses: our forced helplessness, our needs, our naivety. We understand goodness as a condescending attitude towards our natural weaknesses. In this sense, the advantage we will never lose is humanity, that is, the ability to have weaknesses and the ability to support others.

An artificial person of the future may retain these abilities but in some fundamentally new format. Parenthood, mentoring, mercy, empathy - all this will take on hitherto unknown forms for objective reasons. This will happen because we will become significantly less vulnerable and less helpless.

So, to be human means to have weaknesses. This is what makes us social, and this is the only advantage over the artificial humans of the future that we will never lose. Our weaknesses, however, motivate us to cooperate, negotiate, to create communities, to seek friendship, to create families, to love. Perhaps the invulnerable and immortal artificial people of the future will simply scatter throughout space since they will not need each other. Therefore, even if we can become omnipotent, perhaps some of our weaknesses should be preserved.

Weaknesses of a Biological Person

The biological human is beautiful in many ways but also imperfect in many ways. Moreover, this beauty is subjective in some cultural sense, and its imperfection is objective. We admire a person, but all their wonderful qualities, except the ability to survive and evolve with their group, are perceived as virtues only due to cultural conditioning. We have never been the fastest or strongest on Earth, and now we are not the smartest either.

Even though computers have long outplayed humans at chess, we humans still compete in this ancient game. We have taken the computer out of the equation and are choosing a human world champion as if we had come to terms with the fact that we are now playing in a "children's" chess league.

Indeed, our imperfect body and, among other things, our imperfect brain now look like a human cradle. In the same sense in which Tsiolkovsky called the Earth the cradle of humanity, which we will sooner or later leave in search of new worlds. ('Earth is the cradle of humanity, but one cannot live in the cradle forever', from a letter written by Konstantin Tsiolkovsky, 1911.) Russian-born scientist and mathematician Tsiolkovsky is often referred to as the father of astronautics and human spaceflight. So, our body is also a cradle and our main obstacle. To go to other worlds, we need to leave not only the "cradle of the Earth" but also the "cradle of our biological body". If we did not need air, water and food, but only electricity, then we would have probably begun to populate Mars already in the last century. After all, the one-way path was manageable. As for harsh natural conditions, people on Earth also live in less friendly climates.

We have learned to romanticise our weaknesses. How else? We can't live with the feeling of how helpless we are, faced with the forces of nature and chance. Therefore, we are not mortal but merely moving into another world. We are not forced to refuel, but we eat delicious food, we are not deprived of memory, but we love books. There are at least seven reasons that make a biological person weak. This list is probably not complete, but let's list them anyway.

Weakness 1. A biological person is mortal.

All living things are born and die. Due to the change of generations, species of life evolve. Each next generation is, to some extent, a little better than the previous one. And man is no exception here, however, for at least the last two thousand years, man has not become better in a biological sense. Our culture has developed, including medicine, which today allows us to prolong human life, save us from dangerous viruses, and implant organs or limbs that have failed, but this has not made us immortal.

Even though culturally we recognise ourselves as masters of the Universe, in fact, biologically, we are still the same species that is subject to a merciless evolutionary "meat grinder". We are one of many species, but not a single animal dreams of immortality, and we, on the contrary, are sincerely surprised that we will have to die and do not understand what it means to sink into oblivion.

We don't want to die, but how can we get rid of this terrible necessity? We could change our biology and teach the body's cells to regenerate indefinitely. It is theoretically possible to stop the biological clock, although this task still looks like a fantasy. And is it worth it? Our bodies are imperfect for the cosmos not only because they are mortal but, first of all, because they are biological or, in other words, earthly. They are created only for the special conditions of the Earth.

We have long realised ourselves as a soul but have not yet learned to perceive the body as "clothing". Because we are attached to it like children who do not want to sleep without their favourite toy. But suppose we abstract from this "nostalgia" for the biological body. In that case, the option of immortality looks much more realistic and reasonable when a person's "soul" is placed in a non-biological carrier that is not subject to rapid ageing and is also easily replaceable.

A person whose personality is transferred from a biological medium to an artificial one will also be able to feel. The joys of perception themselves, sensations, the so-called qualia, are formed in the brain and not in nerve endings. They do not originate in nerve endings, so an artificial body can be as sensitive or even more sensitive than a biological one.

The acquisition of immortality looks like a completely natural stage in the development of humanity. Many of us already consider our "soul" immortal, but it could become truly immortal only if we intervene in the providence of nature and change the "frame" for the "diamond" of the soul.

Weakness 2. A biological person uses three types of fuel to maintain life. Each type of fuel is critical.

These are air, water and food. Remove any of the three ingredients, and the person will die very quickly: no air after a few minutes, no water after a few days, and no food after a few weeks. However, rather than seeing ourselves as weak, we have created cultural forms of admiration for those weaknesses. We enjoy clean air, and not just to breathe, we quench our thirst, bringing into this process much more than just "refuelling", and we have generally turned food consumption into an entire socio-cultural institution that divides people into classes.

Our creations do not die without fuel, and, moreover, most often use one or two types of fuel, which are much more common in the Universe. Moreover, technology can vary different types of fuel so that if one is not available, then another is used. Try breathing water or eating air. No, we were created to live only where we appeared, where completely unique conditions were created.

Today, we are forced to consider the Earth the only possible place for us because not only can we not fly on our ships to the nearest suitable alternative, "Earth," but we cannot even detect something of the kind in the visible part of the universe with our powerful radio telescopes. It seems that our planet is very special, and if a catastrophe, insignificant by cosmic standards, occurs, all biological life that we know in the Universe will simply die.

Weakness 3. A biological person is susceptible to disease.

If you send a modern person into the past in a time machine, then the bacteria of the past will probably kill him. As happened to archaeologists who discovered the Egyptian pyramids and encountered strains from the distant past. What will happen if we meet other civilisations in space? Most likely, we are not ready to meet biology of a different origin.

Yes, we know how to protect our bodies from bacteria here and now on Earth in 2025. But how will the expedition to Mars, which Elon Musk is already planning, end? Could it be that completely new bacteria have developed there that we are not familiar with? The most surprising thing is that huge investments can be wasted due to one mutation that we do not foresee.

The artificial body will not have problems with bacteria. It will be burdened with other problems, no less serious. But at least it won't be a microscopic bacterium that you can't even see.

We are far from the perfection with which nature produces our bodies. The materials of our body are still too good and cannot be produced artificially. The flexibility and manoeuvrability of the human hand are also unattainable for modern robots, not to mention the body's ability to "heal wounds," carry out complex thermal control, and so on. But it's only a matter of time. And when we have learned to produce artificial bodies, they will be equally beautiful and reliable.

Weakness 4. A biological person is born helpless and takes a very long time to learn.

Above all, we value parents, educators and teachers – all those who teach us to live and survive in this world all those who make us independent and capable of solving complex problems. But the fact that each new generation of people must be trained anew looks like a weakness against the background of advances in machine learning. After all, if AI has mastered some skill, then there is no longer a need to ever teach this skill again. Moreover, this skill can be multiplied (copied) and used everywhere.

In one way or another, skills are neural programs. Sequences of neuronal signals that send signals to motor neurons and regulate our actions. It's touching to watch children train their neural networks by constantly making funny mistakes, falling awkwardly or mixing up words, but this, again, is just a romanticisation of weakness. Without parents and educators, children are constantly in mortal danger because they do not know how to do anything.

But what are our skills, really? And how are they related to what we call the soul? This is an old problem called the mind-body problem; however, presented a little differently. Today, we know that our skills are just neural programs that our brain forms during the learning process, creating the necessary neural connections. And the soul is probably just a volitional impulse of life that moves us, that is, what is usually called "spirit". Therefore, if, for some reason, a child's brain already had all the neural connections of an adult's brain, then the child would be able to think the same way as an adult. And do the same as an adult. He would undoubtedly be weaker than an adult and would not have the same body, but that is the only difference. In the case of an artificial body, this barrier would be overcome, and children would be independent from the first day. It remains to be seen whether this would be wonderful or terrible.

On the one hand, if "newborns" have the skills of adults from the very beginning, we will lose the institution of childhood so dear to us, but on the other hand, this will guarantee that "children" will be able to protect themselves from the tyranny of adults. There will be no childhood traumas, although at the cost of the fact that this will probably destroy the institution of the family in its modern form as unnecessary.

This is sad, but this could be the price if we want to continue to develop our civilisation and populate other worlds. I would like to believe that we are wrong on this point. Perhaps the family as an institution will be preserved for other purposes. The future is likely to contain challenges and problems that can unite us.

Weakness 5. A biological person has significant memory limitations.

Our brain is far from an ideal means of storing information. We are created to remember not data but skills, that is, sequences of symbols or actions. Our memory, just like the memory of artificial intelligence systems, is built on the principle "if A, then B." To remember something (B), we need to give some signal (A) at the input. Moreover, (A) is usually not just a question of what we want to remember.

For example, to remember the second line of a poem, we need to "load" the first line of the poem into our neural network. To remember where we put the keys, we need to reconstruct a sequence of events. The condition "if A then B" is sometimes called an association, but it is more than just an association. This feature is associated with the structure of a neuron, which has input and output signals. To get a certain output signal, you need a corresponding input signal.

This feature of our memory makes it easy to remember sequences of actions, but when it comes to listing unrelated data, our memory is far from ideal. For storing large amounts of data, the structured memory of a regular computer, based on classifications and queries by address, is best suited. In principle, books, encyclopaedias, and any databases perform the same function.

For a long time, we have not used our memory to keep a lot of data in our heads. To do this, we use external media such as books and electronic databases. And this union – our skills, passed on from generation to generation, and external storage of information – has led humanity to incredible achievements. Today, it has become clear that AI systems can also accumulate skills much more efficiently than we can. Machines or robots are potentially capable of not only storing information but also independently using it to solve various problems.

A revolution in this matter will begin when AI learns to independently search for the necessary information in electronic databases. For now, we do it for them, setting up the search. But when AI can learn on its own, collecting all the information in all possible contexts from a data file, we will have a storage of information unprecedented in terms of volume and efficiency. You can ask the electronic assistant not only for some household items or reference information, which is already easy to find on the internet but also in general for any special information that can be obtained by logically deducing it from the entire array of our electronic data.

Weakness 6. A biological person becomes accustomed to cultural concepts and values and then cannot easily change them when necessary.

The peculiarity of our memory, as already mentioned, is that we remember something always in connection with something that reminds us of it. For example, we remember some folk traditions in which we gladly took part in childhood, in connection with our love and attitude towards our parents and other close people. A traumatic experience is also remembered in relation to what was experienced at the same time. And then, even in adulthood, this connection turns out to be not only logical and emotional but also neural. These neural connections remain with us forever unless some events occur that are more striking than how the connection was formed.

We get used to the cultural concepts of our environment, our family and surroundings, the customs of our profession, the values of our class, our party, our way of life, our successes and failures, our acquaintances and the style of our favourite media, TV series, etc., not just in the sense of following traditions and the like. This is how we form the neural connections of our brain, and ultimately, almost all of our memory turns out to be "biased." These neural chains then take part in all computational processes and influence our decisions against our will. We almost never realise that we make decisions under the influence of certain cultural narratives that have become the structures of our neural calculations.

Neuroscientists often talk about brain plasticity. This is the fact that our brain is capable almost of miracles, by changing the functions of individual parts depending on need. For example, if an area of the brain is damaged, then another can take over its functions. This is a truly amazing property; however, this plasticity manifests itself when the brain still has the resources and the potential to form new connections of a certain type. The prototype of Mowgli in real life never learned to speak or even walk on two legs. The moment was lost. If we do not start speaking a language before the age of four, then it is unlikely to become our native language. If we fill our memory with something in our youth, then this something will always be with us because this memory constitutes not only storage, but also all our computing means. It influences the way we think about everything else forever.

If the life of a mature person unexpectedly changes greatly, and he must get used to new traditions, ways of thinking and values, then most likely he will experience very great difficulties for many years, and maybe throughout his entire subsequent life. Such intellectual inertia is, of course, a weakness. People often choose between the good and the familiar in favour of the familiar, between tradition and a rational decision in favour of tradition, and between an unrealisable fantasy from childhood and an understandable benefit in favour of fantasy. All this, unfortunately, makes a person weak.

Weakness 7. A biological person has needs that conflict with the interests of other people. This gives rise to conflicts.

Marx understood conflict as a natural stage of the development process. But he also understood development itself as a struggle. However, Marx wrote about biological man, who initially had an evolutionary impulse to fight for survival, that is, for resources. This property, of course, can be understood as a person's strength because it has brought us to where we are today. Yet man's constant desire to seize as many resources as possible regularly leads to social upheaval, and one of these upheavals has the potential to destroy all people.

Survival of the fittest cannot be considered the norm. You cannot create conditions in which the interests of one group of people lead to the death of another group of people. But such conditions exist in any war, any armed conflict. And they, unfortunately, are still happening on our planet, and there is no reason for them not to continue in the future. And this is an undoubted weakness of a biological person.

Do Robots Do Labour?

What is the role of modern technology, and how does technology participate in human labour? There are two very notable, polar points of view on this. For Marx, technology is a means of production, and it is needed, in general, for human labour. The opposite point of view is that technology has a certain independent trajectory of development, and that it is necessary to consider the evolution of technology as separate from that of the human. This position was first expressed by the French philosopher Gilbert Simondon in his groundbreaking 1958 essay, 'On the Mode of Existence of Technical Objects' (Simondon, 1980).

But what makes these positions opposite? Simondon mentions that up to the modern day, the reality of a technical object was in the background, overshadowed by human labour. The technical object was grasped through human labour, thought of and regarded as its tool, auxiliary means or product. However, for the sake of the good of man himself, it would be necessary to carry out a reversal, which would allow what is human in a technical object to appear directly, without mediation by relation to work. It is labour that must be understood as a phase of technicality [*technicité*] and not technicality as a phase of labour since technicality is a whole, of which labour is a part, and not vice versa. (Simondon, G. (1958), '*Du mode d'existence des objets techniques*', Paris: Aubier, 1989).

Probably, the problem in determining the place of technology is due to the fact that its role has greatly increased, and in some cases, artificial intelligence no longer seems to be a part of production. Simondon was right back in 1958 that labour should be seen not as the basis of fundamentals, but as part of something more global. By the way, how should we understand human labour today? If something was made by artificial intelligence, is it the result of the work of the developer of this artificial intelligence?

According to Marx, labour is a purposeful human activity during which, with the help of the tools of labour, he or she influences nature and uses it to create use values necessary to satisfy needs. That is, for Marx, the following aspects are important in human labour: 1) this is a purposeful activity, 2) this is an activity aimed at satisfying needs, 3) this activity uses the resources of nature, 4) this activity is carried out with the help of technology.

The first point means that labour is carried out consciously and, therefore, cannot be carried out by an unconscious machine. But this is no longer entirely true today if we look at AI. To do something consciously and purposefully for a person means to be motivated, that is, either to make a moral effort or to be motivated by an external factor. For example, this means that the worker can be paid. In addition, purposeful action means acting with intent. But artificial intelligence does not need motivation, neither internal nor external. In addition, artificial intelligence can produce a valuable product by accident without a predetermined plan. In other words, the fact that work is always a purposeful action is already debatable today.

The second point means that labour satisfies needs. Increasingly, however, modern technology does not satisfy but creates needs that enter our lives. This means that labour is needed globally to deal with needs rather than only to satisfy them. Having become a commodity, labour has turned out to be something else.

The third point about using resources of nature is still relevant, but more and more often, people work with information without using natural resources in any way. That is, this point is now also problematic.

Finally, the fourth point about technology was very important to Marx as it described the workers in the factory. The equipment for working at the factory was very expensive, and only a capitalist could buy it. Now the equipment for work may not always be so expensive. For example, a blogger can work with affordable technology. Strictly speaking, work can be done without tools at all. For example, a psychologist hardly uses any devices, but his activity is labour. Marx attaches great importance to technology, making it a key element of labour, but this does not mean that labour is impossible without technical means.

On the issue of the role of technology in labour, we are more on the side of Simondon since Marx's approach to technology was not objective. Marx looked at technology at a specific historical moment and gave it only the meaning that technology had in his time. Today, everything is different. Technology, as exemplified by artificial intelligence, is no longer just a tool that we use like a hammer. Today technology has the ability to surprise us, to do something we never thought of, it can teach us something new. This means that labour must be understood in a new way in modern times as well.

On the other hand, however, Simondon's approach also raises many questions. Saying that labour must be understood as a phase of technicality [*technicité*] he unjustifiably separates technology from the human. In our opinion, in order to look at this situation objectively, it is necessary to see something more global than technology and more global than human. This something global is a certain evolutionary agent that exists in a certain evolutionary environment. Moreover, it can have a variety of manifestations from a cell to a conscious machine. Its main feature is that it persists over time in its new versions – it evolves. The evolutionary agent and the evolutionary environment, in this sense, determine both the meaning of labour and the current role of the human and the technological.

When talking about infinite time and space, about evolution as such, it is very unreasonable to think that the human is only what we see today. It is also unreasonable to look at technology and see it only as it is today. It is important to see why an evolutionary environment arises and how an agent evolves in it. If you look at the problem of the correlation between technology and labour in this way, then it turns out that both technology and labour are the ways through which an agent preserves and develops itself.

We know that nature is the evolutionary environment. It was in it that life and biological evolutionary agents – species – arose. But another evolutionary environment has become a culture in which there is a place for its own evolutionary agents – technology. It is simply no longer possible to consider a person

111

as just part of a species since a person is a person precisely by virtue of belonging to a culture. The 'person', if you like, is a technology that created itself.

At an even more global level, however, it is necessary to rise above both nature and culture and to conclude that only that which is capable of development or evolution is preserved in time. We can assume that the most amazing things arose at different times in various places in space, but most of them simply were not preserved in any form. Moreover, while losses are inevitable under the influence of external factors, conservation means accumulation – the accumulation of useful changes.

In the broadest sense, an evolutionary agent is a group of objects that are produced by the evolutionary environment with some accumulation of useful changes. With this view of the problem, the boundary between species and technology is also gradually blurred. What becomes important is not what evolves but what the mechanism for accumulating useful changes is. It can be biological at the DNA level, social as the transfer of skills from generation to generation, or cultural through the creation and preservation of objects of material and intangible culture – in particular, technology.

For tens of thousands of years, the human has been changing themselves, constructing themselves and, changing the body and the consciousness. And this process has not stopped, since today we continue to constantly change. We create tools, and then they gradually become part of our "bigger" body. For example, most of us have a strong idea that at any given time we can look at our face for some purpose, to fix our hair or check something. Try removing all the mirrors from your house and covering your phone's cameras with tape, and in a very short time, you will feel as if you are missing some important organ. You won't be able to look at yourself, although you're used to it being as easy as shelling peas. But it turns out that desire is not enough for this and that mirrors are also needed.

Sometimes, this automatism can be funny. For example, some children try to control a TV screen with their fingers, one which does not have touch screen technology. It seems strange to them that something isn't working. But this would be even funnier if an adult, accustomed to this technology, tried to do the same with a paper book as if it were a tablet.

In this sense, technology is not divorced from us, but it is not just a tool in the hands of man. We change along with technology, and in order to understand where the border lies, we should probably ask the question: is there something that remains unchanged in a person despite the development of technology? In this case, technology should be understood as literally any instrument. What, really, do we have in common with Neanderthals? What do we have in common with the great ape, which stands even earlier on the path of evolution? But hominids already used tools, and the changes caused by these tools were already there. If you think carefully, modern man is a creature who knows how to use the technology and the cultural tools that he created. That is, we are different today precisely to the extent that our culture and technology, as part of it, are different.

Let's imagine that in the distant future, a person will not have to look for information in books or on the Internet and will instead, simply by thinking about something, immediately access this information mentally. Moreover, this will immediately contain not only the useful information but also an assessment of it. If such technology is developed, it will change not only the way we collect information but also our understanding of who and what we are. Very soon, it will seem to us that this is the norm and that it is natural to be so smart and all-knowing. Today, it seems to us that it is quite natural to move from one continent to another in a few hours. But before, even a lifetime might not have been enough for this, and once upon a time, people knew only very limited spaces around them without even thinking about other continents.

We have much more in common with the technology we create today than with the Neanderthals, not to mention the great apes. If you were asked, what would you choose: to be in an artificial robot body or to return to the cave? Considering how much time we spend today reading social media, watching TV or playing computer games, we hardly use our biological body. And if we didn't have to work to earn food, since robots don't need to eat, then many of us would completely disappear into consuming content. We are undoubtedly closer to robots than to monkeys.

But what about the difficulty, then? If technology allows us to turn into idle couch potatoes, will we not return back to the animal state? Friedrich Engels wrote in 'The Role of Labour in the Transition from Ape to Man' that it was labour that became the key force that created man from ape. So, won't humans soon turn back into monkeys if robots do all the work for us?

In fact, as we have already seen above, in the last century, work was understood differently. And whether work makes us human is also controversial. If work makes us human, then why don't those who work very, very hard feel like human beings anymore? If a person is overloaded with work, then he feels more like a gear that turns other gears than a person. Theoretically, we could train chimpanzees to do what we call work, but would that make them more human? In other words, we return to the fact that it is not the work itself that is important but what it is aimed at. Does the forced labour of an inmate in a prison make him more human?

Man created himself, trying to control and manage the world around him. And labour is just one of the forms of such management in the world. In addition to work, there are other forms of control, for example – this is play, this is cognition, this is meditation, and so on. What makes a person human is what he or she sees as actions leading to results and how he or she perceives his or her impact on the world around them. And the most important thing here is that this awareness then changes the person themselves.

Now let's ask, will the robot's activity be labour? If work is a form of control and not just the execution of work, then this means that work is not just useful actions for which money is paid but an activity that affects both the world and the agents of the labour themselves. It is an activity that becomes a source of causality, just like play, learning or meditation. Let's give an example.

If AI generates some images at the request of a developer or user, then this is not the work of AI. The user selects the necessary queries, gradually adjusting the result, obtaining exactly the image that solves the problem. It depends on the user whether the problem will be solved. He has the ability to use this AI as a tool, and his experience creating queries is essentially programming but in a more advanced form. That is, this work is similar to the work of a programmer, although without a special language.

Now, imagine that three AI systems are working in parallel. One of them generates text queries, the second generates images based on them, and the third, independently of the first two, determines how well the image corresponds to the task and instructs the first AI to adjust the request. In this case, it is not the AI user, but the AI that independently corrects the request, and what the result will be depends only on it. In this case, it will be closer to the concept of work.

There will be sceptics who will say that the initial text request is still sent to the first AI by the user. That is, it still affects the result. It turns out to be a modern version of the homunculus problem, in which no matter how we simplify the user's original task, the request still comes from him or her, and the AI only solves it. But that would be a mistake. The fact is that work does not imply independent goal-setting. It implies precisely its independent achievement, and this corresponds to our second example. Here, the user of a system of three AIs acts as a manager, and the three artificial intelligences together correspond to the employee.

It turns out that one individual AI cannot work, but in the aggregate of several independent AIs, this is already possible. Perhaps our brain is structured in much the same way as a collection of a large number of independent AIs. We receive a task, and then work happens inside our consciousness, which is aimed at finding a solution and the best implementation of the task.

By analogy, one bee or one ant cannot achieve much on its own, but when there are many of them, they can do absolutely wonderful things. For example, they can build an anthill in which everything is very cleverly thought out. The ant's intelligence is not sufficient to control the world around it and transform it. But the intelligence of an anthill is significant.

A group of independent AIs linked together can change the world and change on their own, adapting to new challenges. Such a group can act as a labour agent and maybe in the future as an economic agent, if this makes sense.

For example, some AI developers in the future will be able to create "farms" of groups of independent AIs, giving users the opportunity to use their abilities to make money. At the same time, more successful AI groups that earn more money will be able to buy more advertising for themselves and better promote their services. This way, natural competition between robots can be created, and as a result, AI systems will independently be able to produce a better product without human intervention. Such activity can no doubt be considered labour because it will influence both the world and the agents of labour themselves.

Artificial Person as a Successor to Biological Person

Just a couple of words can be said about this, limiting ourselves to stating the fact that the biological human has almost exhausted his potential. But it is still extremely important to show that the artificial human is indeed our successor. He rightfully is and should be able to count on our support and love. To prove so is the task of the following, perhaps overly extensive but very fundamental, section.

What will we prove? We will prove that there is no natural human. Human is culture. And if culture goes beyond our biological existence, then something else becomes human.

Four Horsemen of the Human

How is a person different from an animal? There are many opinions on this matter. For example, one could say that humans differ from animals in the degree of development of the nervous system and the capacity for complex social behaviour. However, observations of social insects such as wasps and ants show that even tiny animals can behave in complex ways to fulfil social roles. But if we pay attention to the behaviour of mammals, such as mice or, say wolves, we find ourselves in great difficulty. Their social behaviour will not be so dramatically different. At the very least, we can easily find in their behaviour what we consider to be our own, purely human, behaviour – say, a social hierarchy or complex interpersonal relationships.

What, then, is our difference? Walking upright? Speech? Brain size? Work? It seems that the formulation of the question itself is incorrect. Who are we asking about, after all, about the individual or about society? And does it matter? It certainly does. The fact is that outside of his society, a person is not a person. To be human means to be part of certain relationships that we consider human. Mowgli will never become a man in a pack of wolves, and Robinson Crusoe, left alone, is on the verge of humanity and artificially creates human relationships on his desert island to remain human, where it is especially difficult and, most importantly, unnecessary.

Therefore, the real question about the human is a question about human society. And it is a question about how human society differs from animal society. If you pose the question this way, then walking upright, brain size, or even the presence of speech ceases to seem like an answer. The answer must lie in the way society develops. How it accumulates changes that lead to its development. Let's try to figure it out.

How, in fact, do the changes that lead to development accumulate? In biology, this accumulation occurs at the level of changes in the DNA code. Thanks to the mechanisms of natural selection – heredity and variability – beneficial changes are preserved in the genes of more successful representatives of the species and are thus passed on to future generations. Development occurs because each new generation, as it were, makes its own contribution to the biological code of its species. DNA turns out to be the "book" of the biological wisdom of the species, the code of the biological "culture" in which the achievements of the species are recorded.

However, biological evolution cannot create humans alone. The biological evolution of our species, homo sapiens, has been effectively stopped for a long time. And the biological mechanism of accumulation of useful changes has ceased to make a noticeable contribution to the development of humanity. We don't grow wings, our brains barely change in size, and nothing significant happens that could explain the rapid development of humans over the last few thousand years. This means that useful changes were accumulating in some other way during this time.

It can be assumed, and this will not be a risky assumption but rather a commonplace one, that from a certain moment, the mechanism of accumulation of useful changes for human society became non-biological. What then? Social?

No, not social. Or rather, not only social. Since the social type of accumulation of useful changes occurs in both animals and primitive people, respectively. For example, newborn wolves cannot hunt. While playing, a wolf pup can kill a mouse and then starve next to the dead mouse if its parent does not teach it what to do. The ability to hunt among wolves is transmitted not genetically but socially. And what prey the pack will teach him to hunt depends on the habitat. The pack trains puppies to hunt each animal separately. First, they are given the meat of this animal, then the meat with the skin to get used to the smell, etc.

If a pack migrates, and adult wolves are forced to learn how to hunt a new animal, then having acquired this skill, they will then pass it on to the next generation. That is, beneficial changes in animal society may no longer be related to genes. Their accumulation occurs through the social mechanism of learning.

Look even further into the past. You will find that an anthill or a beehive are very complex formations in which not only social relationships take place but also "manufactured" objects – the anthill or beehive itself. Despite the small size of their brains, these insects have accumulated beneficial changes through the social mechanism for millions of years. Not all insects build such complex structures; Therefore, the ants must have learned this. But they couldn't learn it alone, so they did so as a group.

Each step that brings the ant community closer to the ability to build the anthill we know today took, apparently, hundreds of thousands of years. At the same time, mastery of this skill took place according to a social mechanism and its fixation according to a biological one, that is, at the level of genes. Modern ants are born already able to build anthills.

We have come to understand that the accumulation of beneficial changes can consist of homogeneous or heterogeneous stages. Specifically, these are stage (1) creating changes and stage (2) maintaining changes. Some changes can be created, say, at the level of social interaction, as in the example with ants, and then preserved biologically at the genetic level (the formation of an unconditional instinct). Or, as in the example with wolves, in both stages, the creation and preservation of changes can already take place according to the social mechanism (conditioned instinct).

It is important to make one addition here. In the case of a pack of wolves, learning a new type of hunting (that is, creating a change) can take place not only through a social mechanism but also through a mixed psycho-social mechanism. After migration, a wolf can say, learn alone to hunt an animal that is new to it and only then teach this to others. This is important because it shows a more complex mechanism for the accumulation of changes, consisting of three stages:

1. Insight at the individual mental level.
2. Testing the insight at the social level and consolidating it as general knowledge.
3. Preservation of knowledge through transmission from generation to generation.
4. An additional stage is gene correction due to changes in behaviour.

For millions of years, this mechanism (along with the biological one) allowed both animal communities and our ancestors, the anthropoid apes, hominids, to develop and adapt to new conditions. But how, then, does our modern society differ from animals? Animals, like us, are able to accumulate useful changes through a social or psycho-social mechanism, that is, transfer knowledge to new generations and record them in the form of conditioned reflexes. Why are we the only ones who have achieved such noticeable development?

Probably, a person accumulates useful changes in some other way, by some other mechanism, and not just biologically and socially. What mechanism is this?

And here we are, in fact, approaching the purely "human" since we are talking about the cultural mechanism of accumulation of changes. What is it? If the biological mechanism allows for the accumulation of useful changes through gradual changes in DNA, and the social mechanism allows for the accumulation of changes in the learning patterns of young children, then the cultural mechanism implies the accumulation of changes in the form of material and intangible culture. Using an analogy, we can also talk about the complication or enrichment of the cultural code.

If we break this process down into stages, then the three stages discussed above will no longer be enough. The cultural mechanism also involves additional stages of encoding and decoding information through objects, signs and concepts. How does this happen?

Let's imagine such a situation. It's the Stone Age. A young member of the community finds a stone knife belonging to one of the hunters and begins to play with it. A stone knife is a developed technical means of that time, and to produce it our ancestors had to perform very difficult actions. Namely, with the right blows, they chipped off the excess from the stone, and sharp points formed at the site of the chips. Such a knife could be just as sharp as a modern metal one. Taking this knife into his hands, our young homo sapiens has encountered not just a piece of stone but an element of material culture. He could learn many useful skills from the stone knife itself even before other people taught it to him through the social mechanism.

A modern boy or girl who can read is able to learn almost everything from books. That is, without the help of an elder. Undoubtedly, this is much more complicated, and the social form of learning is extremely important today, but nevertheless, we are talking about the fact that since time immemorial, there has been a purely cultural mechanism for the transfer of knowledge and skills. Moreover, the cultural level can always be distinguished in a mixed form of education since it is associated either with artificially created objects or with artificial concepts.

Even if an adult repeatedly shows and explains to a child how to ride a bicycle, the most important things in this matter are still taught to a person by the bicycle itself. As an element of culture, it already contains both the concept of riding and the riding itself.

Let us then highlight the main stages of the cultural mechanism for the accumulation of useful changes:

1. Mental level – Insight, creation of ideas, concepts.
2. Social level – testing the idea.
3. Cultural level – coding of an idea by its embodiment in a material or intangible object of culture or cultural tradition.
4. Cultural level – transfer of experience through the preservation of cultural objects and their decoding in the future.
5. Social level – preservation of this knowledge through transmission from generation to generation.
6. Mental level – decoding cultural objects.
7. Biological level – additional stage: Gene correction is possible due to changes in behaviour.

We see that, according to this scheme, the mental level itself does not have a mechanism for maintaining changes. Changes here persist biologically (7), socially (6) and culturally (5). The psychic level – or consciousness – performs an important but auxiliary function in this scheme. The psyche retains memory for the duration of an individual's life and generates insights, encodes and decodes cultural information. But human life is much shorter than the life of our biological "species", the history of our earthly society, or the history of our earthly culture. So, over the last few thousand years, human society has been accumulating major beneficial changes in the form of culture. Not biologically and not socially. This is what makes us different from animals.

Two and a half thousand years have passed since the time of Socrates, and we can still turn to the primary source of his dialogues – the texts of Plato. Yes, we will listen to a lecture about Socrates, and we will read about him from historians of philosophy, but we will also look into the original cultural monument, and we will read it in ancient Greek. And we will try to understand him, who lived so long ago, exactly as if we were in his time. This is impossible without the original source. The cultural movement forward that we have seen over the past few millennia is a consequence of the fact that we have been able to look back all this time. Look into the past and try to decode the cultural codes of yesterday, a hundred years ago or a thousand years ago. An opportunity that animals do not have.

When we talk about the psyche of an animal, we mean that it does not decode cultural objects. Recognising culture and transmitting it is a human ability. Moreover, an ability that manifests itself only in human society. That is, to be human means to recognise and transmit the culture of human society. And if, say, a pet, such as a cat, knows that its bowl is a person's way of feeding it, then that cat also becomes a little human. She connects with humanity.

A domestic animal, depending on the degree of involvement in cultural symbolic processes invented by man, ceases to be just an animal. The same, but to a lesser extent, applies in general to animals living not in the wild but in the human environment. A person "infects" everything around him with culture. Man creates and transmits culture. And if we want to understand what human society really is, we need to first understand why a person transmits culture. Why does cave painting arise, where does the cult of fire come from, and so on, right up to modern culture?

Let's start by asking: what is cultural coding? How does it happen? What is its mechanism? In our search for answers, we will not resort to standard procedures, since they will not answer our question in exactly the way we need. We would like to look at culture solely from the point of view of how and why it sets us apart from the animal kingdom, how and why it makes us set our alarm clocks and wake up early in the morning to get to work or make breakfast for our children and send them to school.

The drive mechanisms that we observe in the anthill, and which force the ants to perform their functions, are social in nature. But people's motivation is far from this. Animals don't have to force each other to do anything. People are almost always busy with just this. The institution of coercion developed in our society cannot be explained solely by social reasons. It has not evolved in animals for millions of years. The essence of coercion is something else, and this "otherness" has a cultural meaning.

Family and society have completely natural origins. Only culture goes beyond nature. This is exactly how we define it in the broadest sense: extra-natural, artificial. But what is its meaning? Hunting with a sharp stone knife gives a person a huge advantage over his prey. An advantage that was not intended by nature. And the primitive religious cults invented by man gave this society a force of unity and confidence unattainable by any animal. Culture, therefore, is a great advantage, but an advantage based on what is not originally found in nature.

Like everything else, culture is a way of solving problems. And the cultural way of solving problems implies that the tasks themselves should have certain characteristics. For example, there is a social objective to keep a fire burning in a cave because fire is a source of light and heat, fire deters predators, and fire can be used to cook food. However, is it possible to convey all this valuable information to everyone who depends on maintaining the fire? Perhaps it would be more effective to connect the need for fire with something very simple and obvious. For example, you can come up with a fire cult.

Maintaining fire for life is a social task, but maintaining fire in the fulfilment of cult duties to the deity is already a cultural task. Moreover, the first fire on our planet was not produced by man but was taken from nature and preserved. The deification of such a fire, which was not yet controlled by man, seemed natural. But this looks natural for us today from the position of people living in a world of fictitious cultural concepts. For primitive people, the coding of social needs in the form of a cultural concept had to be fully justified based on the realities of that time.

What kind of problem should be such that its solution requires cultural coding? Wolves, for example, teach their pups to hunt without resorting to cultural coding. They do not make sacrifices to the deity to ensure success in the hunt. Man, however, has created hundreds of rituals around himself that have no meanings other than cultural ones. Why? What is the pragmatism of culture?

According to Vygotsky (Vygotsky, 1962), cultural communication creates a space of freedom in relation to other levels – social, biological, etc. For example, a person is capable of deliberately starving themself for some purpose, despite the need for food, exhausting themself with heavy exercise, going without sleep for several days, and doing similar things that are unnatural to our biological nature. Some animals are also able to resist their biological nature for the sake of the social, such as is seen in domestic animals. But no animal can possibly resist its social origins for the sake of cultural values, and here, man seems to be unique.

We know how to resist the opinions of our social environment, or even completely refuse communication, or, on the contrary, use means of persuasion to convey our opinions to the people around us. What does this mean in the language of our narrative? Exactly that, a person can put his cultural principle above a social one.

Even in ancient times, in the Old Testament, Abraham was ready to sacrifice his beloved son Jacob as a sacrifice to his faith in God. But even today, for the sake of cultural concepts created by us, among which are nationality, religion, property, wealth, and much more, we can commit crimes, renounce family ties, and break or establish relationships with entire social in-groups.

Vygotsky, as a psychologist, considered situations where a person had to make a choice. And of particular interest, of course, are situations where making a choice is not easy. Jean Buridan's image of a donkey, dying of hunger, unable to choose between two identical haystacks, was one of those metaphors that Vygotsky puzzled over. How does a person really make a choice? He writes (Vygotsky,1983, p.368):

It is true that throughout history, behaviour is entirely and completely determined by the grouping of stimuli, but the grouping itself, the stimulation itself, is created by man . . . Behind the play of stimuli and reactions, you lose sight of what actually happened: a person's active intervention in the situation, his active role, and his behaviour, which consisted of introducing new stimuli. And this is where the new principle lies, the new peculiar relationship between behaviour and stimulation that we were talking about. By breaking down the operation into parts, you have lost the most important part of it: the unique activity of a person aimed at mastering his own behaviour. To say that the stimulus determined the behaviours in this case is the same as saying that the stick got the fruit for the chimpanzee (in Köhler's experiments). But the stick was driven by the hand, the hand controlled by the brain. The stick was just a tool of the chimpanzee's activity . . . The man, you say, didn't know for a second what he would do, what he would choose. The incentive (the lot that fell) forced him to act in a certain way. But who gave the incentive and the coercive force? This stimulus was driven by a person's hand. It was the person who established in advance the role and function of the stimulus, which by itself could no more determine behaviour than a stick by itself could knock down a fruit. The stimulus was, in this case, an instrument of human activity. That's the point.

This is how a person acts, according to Vygotsky. Let's explain this. In order not to become Buridan's donkey and not to die of hunger, trying to choose between two identical haystacks, a person creates a new incentive for himself. For example, he decides that one of two haystacks looks more pleasant than the other, and then the choice becomes very simple. Vygotsky writes that the stimulus becomes a person's tool, with the help of which he gains the opportunity to control himself. But more importantly for our description, through it, he also gains access to control others. And this, in essence, is what specifically human activity consists of. Oddly enough, it is precisely this – managing others or motivating others to work together – that sets us apart from the animal world.

Below we will talk about this, about why and in what sense people manage each other and motivate each other (solve problems). However, it is important to explain which aspect of management is important to us here. It is fundamental for us that the accumulation of useful changes according to the cultural mechanism does not occur where no one encourages anyone to act. The cultural mechanism of accumulation includes the motivation of other people as part of it.

Someone will say: "By discussing how some people constantly encourage others to do something, we may be missing something. Oh yes! What about personal motivation? It is true that self-motivation exists, but it does not contribute unless it becomes leadership, and it does not seek to create new incentives for other people. Artificially created things and cultural concepts – cultural objects – synchronise the actions of people. These things and concepts are stimuli that do not exist in nature.

Self-motivation for everyone is certainly possible, but it has no cultural meaning if it is not synchronised with the same self-motivation of other people. And if it is synchronous, then the reasons for synchronisation are always social or cultural in nature. If this self-motivation also involves other people, then this is leadership and motivation. So, individual motivation either follows from group motivation or itself forms group motivation. If it is fixated on itself, then it is not interesting to us since useful change does not accumulate here.

Culture acts for a person as the environment in which he or she receives freedom from social, biological and other restrictions, but at the same time, culture itself becomes an even more serious limiter. We cannot be stopped, for example, by our parents, but we are afraid to break the law or disobey God. The opinion of our neighbour is not our decree, but we are powerless before the traditions and customs of our society. A person might not be so afraid of the wrath of his relatives if the fire in the cave goes out, but the wrath of the fire deity is much more terrible. It is impossible to hide from him. It is impossible to avoid his retribution. Cultural retribution, as we know, comes from within in the form of pangs of conscience, fears, vague anxiety, etc.

One awkward question will be asked here. What about art and creativity? What about wall painting? What about stone statues? What about household items? Is culture just a reaction to the growing free will of primitive man? Is it possible to roughly push into this paradigm the flight of imagination of the creator, his revelations and ecstasies of creating something fundamentally new? After all, the artist to this day acts not only as a creator but also as a rebel. In art we often want to see protest. Does art really limit us?

But let's remain calm and try to figure it all out as if we were talking about a distant civilisation lost in space and not about ourselves. Indeed, in this case, we will not have an emotional attachment to the object of study. If we look at the stages of the cultural mechanism of accumulation of useful changes, we will notice that the act of creativity (or insight) is only the first stage of this process. Under certain conditions, the second and third stages can also be attributed to the creative process. But only partly. The formation of a cultural code of creation and its further preservation for future generations implies that the creation is coded as a social or cultural value. Otherwise, it does not outlive its creator.

That is, firstly, the act of creativity is a flight of free imagination. But perhaps ninety-nine out of a hundred such acts of creativity (and maybe 999 out of 1,000) have no impact since they are not stored by any of the mechanisms. Neither socially nor culturally. A "masterpiece" can be born as a revelation and even as a rebellion, but it is always preserved for something or some reason. And the rebellion of the creator no longer plays a decisive role here. What is more important is that the result of the rebellion (or, in other words, the act of creating something new) will be preserved in the future through the mechanism of cultural accumulation.

Secondly, creativity is also an impulse. If we turn again to Vygotsky, it turns out that creativity could be viewed as the creation of new stimuli. The creator creates a "masterpiece", and we have a need to contemplate it, consume it, and own it, which was not there before. But to create a stimulus, it turns out that it is necessary to involve the audience in the process of cultural encoding and decoding. This means that a "masterpiece" has value only in a certain cultural environment. In other words, although creativity is aimed at freedom, it is part of cultural circulation and ultimately fits into our idea of culture as a more complex system of motivation and control.

So, we have concluded that a person is able to control himself and others thanks to cultural coding and the creation of artificial incentives. Therefore, culture, in the broadest sense, is a mechanism for coercing oneself and others. The purpose of this coercion is to gain an advantage or, more simply put, efficiency. Human society is more efficient than a swarm of bees because humans have an advantage over bees. It is we who raise bees, not they who raise us. And the fact that the barbarians captured Rome means that Rome at that moment was not effective enough, and the barbarians were. That is, when we say that culture is a form of effectiveness, what we mean by effectiveness is the ability to successfully solve problems. We might even say that beauty is a form of efficiency in the very sense that beauty is effective.

To describe culture in the broadest sense as a way of solving problems, note that there are two broad approaches to problem-solving by conscious agents:

A. Solving problems alone (individually)

B. Collective problem-solving

For conscious agents, Approach (A) belongs to the field of psychology, and Approach (B) belongs to society. However, as we have already discussed, when we talk about society, we mean the cultural environment in which cultural agents and their environment take place. Therefore, approach (B) is a cultural approach to problem solving. Let's give an example.

Let's imagine that an excerpt from Mozart's *Magic Flute* was broadcast on the radio, and you experienced an irresistible desire to go to the opera. However, you have two children, and there is no one to leave them with except your spouse. This problem can be solved in the field of psychology (A) by convincing yourself that going to the opera is worth postponing because of the children. But, fortunately, there is a cultural way to solve this problem (B). You can persuade your spouse to stay at home while you go to the opera. Or, you can persuade a friend or relative to look after your children, or hire a babysitter.

But for this it is necessary to ensure influence on a third party. This could be anything ranging from financial incentives or promises to threats (hypothetically) or a sob story – a compelling cultural concept that would sound convincing to a third party.

Thanks to AI, a third approach is emerging. It is also cultural, but we have not yet fully realised this. There are already robots that can drive a car, but there are not yet robots that can stay home in the evening and look after the children. But they would be much more useful to all of us! The company that produces such a robot and convinces the government that it is legal and safe for children to be left with it will make an absolute fortune. Which brings us to the third approach:

C. Solving problems with AI

In fact, the third approach is still the same as the second approach. Because an AI that is capable of being responsible for children is a moral agent. And even if it is not afraid of losing a job or going to prison, it is still no longer a tool, not an object. Today, AI already has a transitional status from object to subject, and in a couple of decades, there will be no doubt that AI can have a personality.

The difference between approach C and B is not that C is a robot and B is a human. The main difference here is that in option C there is no need to force, ask and convince the AI to help us. The person on the other side, or the robot, becomes secondary. And with the invention of conscious robots and the development of artificial bodies for people, this border will no longer exist at all. That's why, speaking about a cultural approach to problem-solving, we could very well talk about the following sequence of ways to motivate people:

1. A person forces others to do what he needs.
2. A person convinces others to do what he needs.
3. A person teaches others to do what he needs.
4. A person creates an ideology of work.

But after this, we must add a fifth method, no longer requiring incentives but simply a way of solving problems:

5. AI performs all tasks that it can perform on instructions from a human.

Looking at these five points, we first think that this is just some technical classification of the way something could be done in a community of people (and robots). However, let's take a closer look at this classification. Will there be something more hidden behind it? For example, let's ask the following questions:

- When did a person learn to force others?

- When and why did a person learn to convince others?
- When did a person begin to systematically teach others?
- When and why did ideology appear?
- When did AI originate?

Is it starting to seem to you that there is some kind of logic hidden in this simple sequence? Indeed, there are two criteria that are implicitly discussed here. The first criterion is desire. Does the person want to solve the problem? The second criterion is knowledge – does a person know how this problem can be solved?

If a person knows how to do something but does not want to do it, then he should be forced or convinced. And if he doesn't know how to solve a problem, then he should be taught how to do it or inspired to solve it.

The tasks that people face have always been very difficult. One should not think that in prehistoric times and in antiquity, man always knew how to solve this or that problem. There is something significantly different here, namely that we are not talking about individual people but about society as a whole. This is not about the fact that any individual person did not want or did not know how to solve some problem, but about what most people did not want or did not know. That is, we are talking about managing people and at what level certain decisions were made.

The following time dependence can be derived. Initially, at the dawn of civilisation, those who controlled people worked to a greater extent with the desires of people, creating conditions in which the individual desires of a person were subject to the decision of a higher authority. All decisions were made by managers. Work with people's knowledge was done to a lesser extent since the person was not required to make decisions.

With the development of culture and the person himself, the situation changed, and gradually, managers were forced to delegate many decisions to people. This was connected, among other things, with the development of technology since managing technology requires special knowledge. Therefore, the management gradually concentrated on making ever wider sections of the population ready for responsible activities. However, this did not completely eliminate the need to work with the desires of people, and the institution of coercion of people continues to this day. Only at the level of AI did a subject first arise that can be trained accurately enough, but where there is no need to force it.

It is very significant that in any particular era, we can identify one main collective way of solving problems that dominated. This problem-solving method determined a lot, from the state and social structure to the type of art that was produced. In addition, each era corresponded to a characteristic type of leadership, the most important for that time, and a characteristic type of division of people into groups or even, one might say, classes. Let us give a brief description of each such Age, starting with the dominant type of problem-solving and listing its main characteristics.

Table 4. Ages by problem-solving method.

Age	Type of leader	Division into groups	Type of culture and art	Type of problem-solving
Age of Coercion	Leader/ king/ aristocrat.	Class society.	Culture of coercion	The leader makes his own decisions and coerces others himself or with the help of the aristocracy.
Age of Persuasion	Spiritual leader/ philosopher/ artist/ speaker.	A division by religion/ philosophy is added.	Culture of persuasion	The spiritual leader makes decisions and does not force but convinces; for example, he places care of the soul and responsibility on the person himself.
Age of Enlightenment	Educational leader/ scientist/ teacher.	Adding divisions by level of education and specialisation of knowledge.	Culture of enlightenment	The leader of education creates conditions for mass education and acquisition of knowledge. His solutions are embedded in the structure of knowledge, but students can develop it.
Age of Inspiration	Visionary/ party leader.	A division into political parties is added.	Culture of inspiration	The visionary creates an ideology that allows the widest segments of the population to receive and produce the necessary knowledge. The visionary insists that decisions must be made by everyone for themselves.
Age of Artificial Skills and Artificial Man	Artificial Intelligence and then Artificial Man.	Division according to the degree of connection with AI into clusters.	Culture of artificial skills	Artificial intelligence and then artificial humans make the transition to a new type of collective problem-solving. This is associated with obtaining knowledge that is not available to a biological person through "dark science" and other tools.

When you look at a table like this, you can't help but be surprised. After all, we were just thinking about how a group of people can solve problems. We did this theoretically, based on logic and two criteria: desire (motivation) and knowledge (skills). From there, we thought about how decisions should have been made, and that's about it. But what happened goes far beyond this simple logic.

What do we see in this table? The Age of Coercion highlights natural leaders who make decisions and force others to carry them out. Gradually, the leaders form an aristocracy class as

145

an auxiliary class that helps to exercise coercion over more people. The institution of leadership develops and ceases to be natural. That is, it ceases to depend on the leadership qualities of the leader. In this case, leadership becomes hereditary, as well as the rights and possessions of aristocrats.

At the next stage, in the Age of Persuasion, spiritual leaders emerge who can combine this spirituality with conventional leadership. But their function is fundamentally different from ordinary leaders. Spiritual leaders solve problems more effectively because they create and transmit certain values to people. Values are a hidden form of coercion. Therefore, having perceived the value, a person no longer resists spiritual leadership. The values also contain the minimum knowledge that people need to solve community problems.

With the development of culture and technology, the amount of required minimum knowledge increases sharply and cannot be transmitted only through values. In the Age of Enlightenment, educational leaders create the conditions for mass learning and education of the population. An educated person must already make many decisions on his own, so he is often forced to stand in opposition to spiritual leaders. For example, a doctor is forced to make decisions on which a person's life depends, so he begins to reject belief in the supernatural forces of nature.

The further development of social consciousness leads to another crisis in man's worldview, and it turns out that even science cannot answer many questions. A crisis of the foundations of knowledge arises, which shows that leaders at all levels can make mistakes, and there is no absolute knowledge. There is a need for new leaders who impose neither knowledge nor desire – Visionaries. Visionaries create ideologies, that is, concentrated knowledge that allows a person to independently make decisions regarding complex cultural objects and processes. Ideological divisions lead to divisions into political parties. Thus begins the Age of Inspiration, and parties become new social classes.

The emergence of AI was not accidental since it logically follows from everything that happened before. AI does not need to be forced, and it processes huge amounts of information. That is, it consumes knowledge. In addition, AI creates new knowledge. The new era, which is already beginning now, at the first stage becomes the Age of artificial skills. AI can already do a lot of things, but very soon, the number of AI skills will grow exponentially. The number of its skills will many times exceed what a biological person can do in just a few decades. But at the same time, the AI will continue to remain in the position of an unconscious robot that has no real feelings, no motivation, and no desires.

At the stage of the Age of Artificial Skills, "dark science" will arise, which is described later in this book. This is a science that is developed by AI without the active participation of humans. This will allow us to make a new technical revolution and move to the next stage. The next stage of this era will be associated with two events: the emergence of conscious robots and the emergence of an artificial human form.

The question of art in different eras is discussed below in the section on the psychologism of culture.

Divergence between Cultural Values and Public Good

So, culture arises as a response to the development of the will of primitive man. Culture creates an environment for the realisation of creative activity, which is a psychological factor. However, because of creativity, human society gains advantages over nature, over other animals. And the cultural way of solving problems is reinforced. But culture is closed in on itself, and the cultural way of solving problems goes far beyond the limits of natural good, creating its own special artificial good.

Can anyone complain that culture today has accumulated few tangible and intangible values? It is unlikely. Usually, they complain about something else: the fact that a person is antisocial. And this is not surprising since the artificial values by which we largely live are an ersatz of natural values. Let us consider this with an example which we have already mentioned, but which is worthy of full study. This is the concept of goodness.

Ask any person if good is good. And anyone will answer that it's good. But let's figure out what concepts are associated with this word, which of them are natural and which are artificial. And is this always a good thing?

The first ancient concept of good is economic. It is understandable to people; it is understandable to animals too. Goodness is associated with the good and the satisfaction of needs. Essentially, good, in this sense, is something material or immaterial that is necessary for life, safety and pleasure. For example, this is food or shelter, or a kind attitude. Good is also called material condition or wealth. And a person who has gained weight becomes fatter.

This concept is economic, but it is rooted directly in our biology. Good in this sense is warmth, satiety, health and life, as gradations of biological satisfaction and prosperity. The caress of a she-wolf is good. Training wolf cubs to hunt is good. People kill and eat cows, pigs, and chickens that live on their farms and with which their children play, and this is also good in a biological and economic sense. But sometimes, it seems to us that there is something wrong with this because we look at it from the side of later and more complex concepts of good.

Such free use of the concept of good is an example of the creation of a cultural concept. But we are not yet ready to give any assessment of this, so we move on to the second – the social concept of good. Let us only note that love, among other things, is a very important mechanism for expanding the individual to the scale of the family. Therefore, when we talk about the first (biological) concept of good, it applies in many ways equally to, for example, a person and his family. That is, the family is not yet a group but still an individual level of the "group". Although "divorce" is not a rare occurrence for animal families, one can apparently speak of some mixed form of individuality for different periods of time. We do not undertake to define exactly what a family is, considering all the nuances of this complex phenomenon, but society, or meaning, is a group larger than a family. And good for society has a different meaning than for an individual or family.

The social concept of goodness is ethical in nature. In this sense, goodness loses its individualistic meaning and acquires social significance, going not only beyond the needs of the individual but also beyond the family as an intermediate biosocial institution. Good here is still associated with satiety, health and life, but now not for the individual, and not for his family, but for society.

The continuation of the life of the group (considering its transformations) is a condition for the accumulation of useful changes in the social mechanism. These changes are passed on from generation to generation and provide groups with some advantages over other groups. The social concept of goodness includes the division into good and bad actions. And consequently, it implies some social ways of motivating good deeds and preventing bad ones. In this sense, a good deed is an act that helps the group survive, accumulate common good, etc.

It is important to understand that social wealth or good is not an object but a relation between objects. And there are relationships and actions. Let's say grain is a benefit to society, but its essence is in the cultivation of grain and in its distribution, and not in itself. You can have a barn full of grain but not distribute it within a group. The social aspect of the good, therefore, lies not in the grain itself but in the efficiency of its cultivation and the fairness of distribution. Thus, social good is rooted in two issues: the need to act efficiently and the need to act fairly. Anyone who acts ineffectively or unfairly acts against the social good.

The ideal of a social good is a certain reasonable standard of efficiency and fairness in the distribution of benefits, labour costs and responsibilities. A conflict between efficiency and justice is possible, that is, a conflict between cultural and social principles. Often, for the sake of efficiency, a person is ready to commit injustice if this is a means of survival. Less often, for the sake of justice, we can sacrifice efficiency, but for this, society must have a stock of goods or welfare. Otherwise, inefficiency can lead to the death of the community. It is very difficult to say what is more important: efficiency or fairness. How can you think about justice if inefficiency can destroy everyone? However, injustice demotivates people and leads to inefficiency in the very near future.

For example, here is a question: which sin is more terrible: murder, or greed and laziness? You may be surprised at first, but greed and laziness become the cause of death disproportionately more often than murder. This will be a very rough approximation, but in 2021 more than nine million people died from hunger in the world, while less than five hundred thousand were violently killed in the same year on the planet. People died not because of their own greed and laziness but because of the greed and laziness of others, those who either did not want to help or were too lazy to anticipate hunger, etc. And this is even worse than murder since it has no purpose; unless someone is deliberately attempting to limit the population of our planet, but this would only explain greed, and not laziness.

The lazy (ineffective) are killed by nature. Being biologically animals, we are still forced to participate in the cruel process of survival and natural selection, which, although partially corrected by civilisation, has not become any less cruel. Try to be lazy and neglect your needs for safety or food, and serious illness or death will not keep you waiting.

A person is constantly forced to think about the future, about keeping the fire going; otherwise, the cold will kill them. About planting wheat or rice on time. About finding a job and not losing it. About not ending up in a dark alley, not accidentally leaving the door to your home open, not getting caught in the web of a skilled swindler, and not losing your property and your home. Idle, careless people, even if they are initially rich, quickly find themselves in biological millstones. And the consequences of collective laziness lead to great injustices and numerous victims.

True, a lazy person can try to find someone who will work for him. And this is how injustice is born. More precisely, this creates an imbalance of responsibility, which can subsequently lead to the most disastrous consequences.

Using his own example, everyone can evaluate how much effort it would cost him or what it is already worth to support the young and elderly members of his family as dependents. But this is justified by the fact that young people are preparing for their turn to lead an adult life, and the elderly, for the most part, have already done what they could. It is another matter to support a healthy adult who does not want to take part in running the household. What if there is not one such person in the family but two, and so on? Sooner or later, such an unfair distribution of responsibility in the family must lead to conflict. The same thing happens when some part of society becomes virtually dependent on everyone else.

We are talking here about inequality between people, but not in the sense of division into rich and poor, noble and simple, but in the sense of unequal distribution of privileges and responsibilities and, therefore, of the useful efforts that people make during their lives. And this form of inequality is much more dangerous than any other. It is this that, sooner or later, leads to the most disastrous consequences. Wealth associated with great responsibility does not raise such questions in anyone as idle and careless wealth.

Wealth is not socially unjust. It is also necessary for any economy, as is the presence of a means of exchange in general. It is unfair when wealth does not impose on a person a responsibility to society commensurate with this wealth. For example, John Rawls concluded that simple agreement between the government and the people does not guarantee justice. In his opinion, people are interested in increasing their own and decreasing the overall share of benefits in any form of cooperation. To overcome this, principles of justice are needed that would distribute shares of benefits in cooperation on a principle acceptable to everyone. It is necessary to give advantages to the least successful members of society at the expense of the most successful. Let's say, introduce a progressive tax for the rich, through which to support those who are less fortunate. And similar measures.

Taxes are, of course, the most efficient method of allocating resources today. However, one cannot help but notice that this is a fight against the consequences and not against the causes of injustice. And, on the other hand, this is also a decrease in efficiency since abundant social support in social societies can reduce people's motivation to achieve.

As you can see, the social conflict between justice and efficiency is not a simple one. One of the sad consequences of this conflict is that, in the opinion of many, both thousands of years ago and still today, it is easier (more effective) to take away than to earn. But what can be opposed to this, because not only criminals do this, but also any government. Social good, or public good, is controversial, and we will touch on this issue further below.

Artificial goods, which belong to the sphere of culture, as we noted above, are accumulated to obtain a significant advantage over nature. That is, culture is aimed at efficiency, and it creates a kind of artificial balance between efficiency and fairness. A stone knife, or even more so a spear, makes a person noticeably stronger and fundamentally reduces the chances of being killed when hunting a mammoth. And if the leader sends hunters on this risky enterprise, then he seems not so unfair. After all, hunters have spears, and the risk of dying is much less.

And this is where, in fact, the divergence between social and cultural good begins. The fact is that the leader judges the risk reduction based on his assessment, but this assessment may be inaccurate. A spear really reduces the risk when hunting, but in addition to the spear, he can also rely on the power of a magic amulet, a spell or the foresight of a shaman. And we can't blame the leader since he really believes in the power of the amulet and spell. The mechanism that allowed superstitions to become entrenched in tribal tradition is to be blamed.

Culture creates its own values and its own benefits, however, we do not always understand their true value. And, unfortunately, we are talking not only about superstitions.

The point is that biological and social goodness lies in the life of an individual or society as a whole. Therefore, the opposite of such good is death. Again, the individual or society. Cultural good is artificial. It is contained in the life of objects of material and intangible culture. Evil for this dimension of good will be the death of these objects. For material culture, this is the death of things – works of art and architecture, as well as the death, for example, of languages, etc. For non-material culture, this is the death of traditions, values, knowledge and ideals. Within the framework of culture, this is precisely the highest evil. Within the framework of culture, a person is needed to broadcast and multiply culture, but his life turns out to be cheaper than "masterpieces" and traditions.

People have created an environment in which man gains an advantage over nature, but ironically, it is in this environment that human life has ceased to be the main value, not declaratively. Finally, it is at the cultural level that a special concept of "good" appears, namely "a good person." As we remember, at the biological level, it was simply "good." At the social level, a thing has evolved into an act – a "good deed." At the same time, anyone could perform a good deed without becoming good or evil since we were talking about efficiency and justice. At the cultural level, good and evil are personified. He who is good follows certain traditions, shares certain values, has certain knowledge and strives for certain ideals. Those who do not follow and do not share these are evil, and it no longer matters whether they do good deeds or not.

If a person kills for a piece of bread so as not to die of hunger, he behaves like an animal. If a person kills to protect his group from attack by barbarians, then this is a social behaviour. If a person kills someone who shares other values, professes a different faith, different ideals, then this is cultural behaviour. No matter how it sounds.

Let's try to look at the conflict between efficiency and justice more systematically. Everything that happens in the world is usually an interaction or, more simply put, an exchange. For example, in the world, there is a constant physical exchange of heat and mass, and at the physical level, it is always "fair". Physical laws are initially fair – just remember Newton's Third Law on the equality of action and reaction forces. We will not reveal a big secret if we say that physical laws precisely consist in the fact that we fix some "fairness" in physical exchange. True, physicists call it something different, for example, symmetry.

Symmetry is fundamental. Symmetry does not require reasons since it arises precisely due to the absence of reasons, that is, statistically. If an asymmetry arises somewhere, then the physicist begins to think about why this happened, what caused it, etc.

In living nature, everything is also built on metabolism, for example, on the metabolism in the body. To some extent, metabolism is even the definition of what life is. This is, at a minimum, the most important criterion of life and, at a maximum, the very form of its embodiment. Any biological or social interaction can also be considered a form of exchange. We usually judge such interaction by whether there is an exchange of substances, objects, exchange at the level of communication (exchange of signals, information), etc.

The peculiarity of exchange among living beings is that they always strive to make it more profitable (effective) for themselves than for the other party. And the key role here is, of course, played by consciousness and intelligence. It is much better for those species that have a more powerful analytical tool and greater motivation to tip the scales in their favour. For example, predators have learned to attack their larger prey in groups, while their victims usually defend themselves alone or in small groups. This gives predators a clear advantage and violates fair exchange. However, it makes hunting much more efficient.

We can say that nature, in this case, achieves a certain balance between the population of predators and other species. However, this balance is not initially fair. But we still see in it some higher level of biological justice, since balance is necessary for all species. If there were too many antelopes, they would eat all the grass and destroy the ecosystem. But it is important to understand that this kind of balance is not the original, but a derivative form of justice. And nature created it, giving some an advantage over others.

When we talk about homo sapiens, we must have in mind an even more abstract form of justice that arises in connection with man's intelligence and his immense inclination towards efficiency. Human systematically carries out an unequal exchange with nature, consuming but not replenishing nature.

For example, thanks to human activity, some species of animals on Earth were completely exterminated.

Do words like quagga and thylacine mean anything to most humans? Probably not anymore. We destroyed them. In addition to them, man has destroyed the population of the black rhinoceros, northern white rhinoceros, orange toad, Zanzibar leopard and many others in just the last couple of hundred years.

We do the same with minerals. This exchange with nature cannot be called fair because we do not give anything in return. And many of these fossils themselves are not restored.

Yes, it is unfair, however, it makes a person very effective. Without this, we would not be who we are today. The only question is how long this can continue. The peculiarity of an unfair exchange is that it depletes one of the parties to this exchange. We are depleting nature, without which we cannot exist. This was unimportant when man consumed only a little stone, wood and animals, but today, the issue is completely different. After all, we have reached the level where we are, in principle, able to deprive ourselves of clean air, drinking water and the Earth's natural protection from solar radiation.

What should we do? And why do we expose ourselves to such danger? Our human form of justice, our human concept of development, lies in an abstract formula – the achievement of the public good. By society, of course, we mean a society of people. This is a purely cultural concept based on a lot of assumptions, one of which is that human society is, for some reason, more important than black rhino society.

We have not had to justify the killing of animals before since we, by inertia, took this concept from our animal ancestors. Previously, we did not bother to justify the use of forests. At some point, however, we thought that we were only taking and not giving anything in return. The point is not that we felt ashamed but that this was the first time we realised it, and it became the subject of our reflection. This happened back in archaic times, and in response to such reflection, a cultural concept arose that we are in a kind of supra-system of exchange. This is a kind of analogue of nature, which creates a balance of species and all living things, and we also have the concept of balance at a higher abstract level. The balance of material and immaterial (spiritual) is controlled by deities.

In this area, concepts have appeared that reflect our right to take since it is as if we are giving something in return. But we took materially and gave immaterially. Although there is also material, but useless from the point of view of real exchange, compensation for what was taken. This is a sacrifice. Having no material meaning but still real compensation to the gods. Gradually, however, our concepts took on increasingly intangible forms. For example, ritual and prayer in general. Or piety and following the rules of the supra-system (in the sense of supra-natural, spiritual).

Why can I take from nature, from the gods? Because nature is part of a larger world, and I follow the rules of this larger world. I give to the deities, and in return, I take from nature. Culture thus turns out to be a kind of artificial justification for such unfair exchange. But in fairness, nevertheless, we note that all these "atrocities" were created and created by man within the framework of a special form of exchange, which we would call survival.

An extreme form of exchange, in which one of the parties is at a deliberate disadvantage, is an exchange for life. For example, a predator will not survive if it does not catch enough food while hunting. The same thing awaits almost every one of us today if we are deprived of our usual living conditions and sent to survive in the wild. For us, losing our home is almost tantamount to dying. In such cases, the exchange occurs according to the "all or nothing" principle. That is, it essentially degenerates.

In survival, the conflict between efficiency and justice reaches its peak. The "surviving" side strives for maximum efficiency at any cost. That is, justice is completely or almost completely ignored.

What is important here? Survival can be of two types – natural and cultural. And if the first leads to real death, then the second to an imaginary one. We are not so interested in the natural situation when a hungry wolf, in order to survive, "sinks" into eating beetles, lizards or even berries. We are more interested in the possibility of an artificial situation when the problem of "survival" turns out to be not natural but conceptual, introduced at the cultural level. Such a problem can be an imaginary threat or an imaginary enemy, or, on the contrary, an imaginary benefit, for which it is necessary to take interest-free measures.

The need for survival can, in principle, justify any unfair exchange. For example, we can justify the need to draw on the Earth's non-renewable resources and process them, polluting the air and water, by arguing that otherwise there will be a serious reduction in industrial production, which in turn will collapse the economy, standard of living, and throw us far back in development.

How to assess such risks? Do these risks really exist on both sides? What should be the balance between these factors, and is a compromise possible? In fact, such concepts as "development", "standard of living", etc., on the one hand, and "ecological conditions" on the other, are just concepts. And nothing is less reliable than comparing the conclusions of two concepts with each other. But since we are talking about survival, we have a completely reliable criterion that we could use.

This is the question: will humanity have time to find a new home in space before the old one is hopelessly exhausted and unsuitable for life? Or will humans have time to change and become artificial before our world becomes unsuitable for biological man? We don't know. But if you look at this problem this way, it will become clear that there are two main factors or, one might say, two curves in time. One of them is our technical development, and the second is the level of ecology.

Today, these curves are in antiphase, and the higher the technical development, the worse the environmental level. On the other hand, the higher the technical level, the greater the hope of finding a new refuge or, as an important option, the option appears to learn how to prevent environmental consequences from pollution. If you do not develop the technical level, then this opportunity disappears. In other words, to survive sometimes you need to take risks.

Any cultural concept that we create for the sake of survival is a kind of loan that we take from nature. And when the time comes to pay it back, a person must create a new concept and take out a new loan. This can be seen especially well in the example of another sphere of exchange – the social.

People actively exchange their time and things with each other, although outwardly, it may seem that the situation is more complicated. However, any obligations and complex relationships can in fact be expressed in time and things. But, as we mentioned, man tends to gain an advantage in any exchange, so various cultural concepts have been created in which time and things participate in a different capacity. Some say that financial derivatives create an economic bubble, but for humans, creating such bubbles is the basis of the economy from the very beginning.

In fact, any cultural invention can be considered such a bubble. And the main bubble is power. Naturally, it does not exist in nature apart from force. It is power that is the artificial regulator that creates a pseudo-balance of efficiency and justice, constantly pushing a person to new achievements. Let's look at this in more detail.

The transition from the "flock" type of social organisation to the "tribe" type is associated with the emergence of leadership. If the pack is controlled by instincts, which unite into a kind of collective mind of the pack, then the tribe already has a leader. Leadership has a social nature and is a basic social value. However, leadership arises in response to challenges and problems faced by society. And leadership comes naturally in the process of solving these problems. However, a cultural tradition of leadership and power in general develops, which continues to operate subsequently in the absence of obvious external challenges.

A leader who helps his group achieve important goals, including survival goals, grows from a leader who has proven himself situationally into an ancestor, a leader. This transformation means a transition from social to cultural leadership. The leader deserves his authority, and he is supported only as much as it benefits the group. The tribal leader becomes a cultural concept, a concept of power. Things appear that symbolise power in itself. If this is a military leader, then his power could be symbolised by a special headdress or decoration. If the leader was a priest and his power was sacred, then it could also be personified by an object necessary for sacred rituals.

It is known that power did not immediately become hereditary, and the leaders did not immediately come to the point of using the wealth of the community for their own enrichment. However, both happened. Power itself, in isolation from a specific person, began to acquire cultural attributes, things, and concepts. Until now, the cultural ritual of crowning the kingdom is accompanied by the transfer of objects – symbols of power – from the previous ruler to the new one. This may seem strange in the modern world, but the modern world is only more complex than the archaic one, while the mechanisms of cultural accumulation of changes remain important. Feeling the crown of his ancestors on his forehead, the ruler receives a special "message" from the past, accessible only to him.

Concepts of power are consolidated in the sociocultural institutions that we wrote about above. These concepts are variable and therefore not mandatory; however, they give society various very important advantages over nature. Throughout our history, some sociocultural institutions have turned out to be more important than others, and we can see how this happened against the background of different forms of political leadership. And also, against the backdrop of a conflict between efficiency and justice in the struggle of societies for survival.

The original form of justice given to us by nature, as stated, is the balance of the coexistence of species expressed in the food chain. There is no hint of any kind of equality here, as one might naively believe. We still must explain to children that there is nothing wrong with eating animals. We also shamelessly appeal to other natural "values" as to some supposedly fundamental attitudes. Animal savagery implies the right of the strongest, the right to kill, and, generally, the absence of any kind of equality. Therefore, today, the appeal to natural values looks like manipulation. A person easily abandons everything natural if it gives advantages to society.

Primitive society recognises the value of leadership as the basis for survival and prosperity. Leadership has many sides, but the main one for early cultures is the ability to coerce. Coercion, in turn, is institutionalised in traditions and customs. Coming out of primitive society, we have people with a clear idea of power – military, sacred or based on eldership. But we cannot say that primitive society is a coercive society, since leadership in this society is still very strongly connected with the personality of the leader. The connection between the leader and the community is still very close; every member of the community knows the leader, and theoretically, the leader can be overthrown by any member of the community.

The efficiency of society currently turns out to be higher the more people's efforts are directed towards the same thing. Today, we call this division of labour. Therefore, the tradition includes the concept of such justice as the justice of compulsion to certain useful activities. Justice is fixed in the customs of submission to the will of the leader. For example, it is the custom to carry out work as directed or the custom to give a significant part of the wealth to the leader. Today, we conduct professional activities, and today, we pay taxes. The traditions of coercion turned out to be very important for the development of our society.

The institution of coercion, however, reaches its full flowering not in primitive but in slave-owning (unfortunate name) society since coercion becomes the professional activity not of the leader or pharaoh himself but of the intermediate class (the aristocracy). This form of work is being formed as coercion, what we call management today. Leadership in a slave-owning society is detached from the people since members of society often no longer know their leader. He is now separated from society by a layer of managers, from whom the aristocracy class is subsequently formed.

A three-layer society emerges, where the main leader remains in the position of leader but only in relation to his aristocracy. It is the members of this layer that still can overthrow his power. The rest of society, in relation to the leader, is in a position of subordination with extremely limited ability to influence their position. Therefore, the conflict between justice and efficiency in such a society takes place at two levels, between the king and the aristocracy and between the aristocracy and everyone else.

Efficiency in such a society will be higher the more successfully the aristocracy forces others to join in common activities. This leads to the fact that the aristocracy becomes stronger and begins to exert a strong influence on the leader. At the same time, the management structure becomes significantly more complicated. A special form of tradition arises, which is consolidated in the form of general cultural norms or laws. If community members independently preserve and defend the traditions, then the preservation and control over compliance with laws also becomes a special form of activity.

We usually think that laws are a social norm, but this is far from the truth. Laws appear not to maintain justice in society but to increase the effectiveness of coercion. In other words, laws are another form of culture through which they artificially restore the balance of efficiency and justice. Coercion shifts this balance in favour of efficiency, and the law artificially restores it.

For example, 'The Code of Hammurabi' (Translated by L. W. King) is very indicative in this sense. Probably one of the main problems in Babylon was slander, which fits well with our idea of Babylon as a collection of numerous peoples and as an important trading centre. Slander placed a burden on the court, so the Laws of Hammurabi paid great attention to the rules regarding slanderers. At least, that's what it seems at first glance.

Paragraph one says:

> 1. If anyone ensnare another, putting a ban upon him, but he cannot prove it, then he that ensnared him shall be put to death.

Several more paragraphs of the Law are devoted to this. For example, eleven:

> 11. If the owner does not bring witnesses to identify the lost article, he is an evil-doer, he has traduced and shall be put to death.

With some additional analysis, we will still decide that these points are directed not so much against libel but against those truly injured plaintiffs who do not have the opportunity to prove the guilt of the defendants. Is this fair? Not at all, but these initial norms were supposed to significantly reduce the number of appeals to the courts, that is, increase the efficiency of the court. After all, proceedings in cases in which it is very difficult to prove guilt take a lot of time and do not lead to anything good.

The motive of the lawmaker, understandable to us, however, could not be understood by the contemporaries of the law. For them, these norms became the norms for the fight against slander and justice itself for several centuries. They could reason like this: if a person slanders another and wishes for his death, then he himself deserves death. And this is supposedly fair. The legislator here, on the contrary, did not care whether the plaintiff was slandering or simply could not prove his case. Both cases for the legislator are a waste of the judicial officer's time.

Laws have retained this spirit to this day. For example, in Russian criminal law there is a rule on the sufficiency of evidence of guilt. All evidence must be reviewed for sufficiency, considering the importance of the decisions being made. This sufficiency is determined scientifically and is a very important criterion for making a verdict. However, the very concept of "sufficiency", with an external desire for justice, in fact, performs a completely different function.

Indeed, the investigation and determination of guilt cannot last forever. This is a job like any other and must be done effectively. Therefore, there must be criteria for the sufficiency of guilt that can be publicly demonstrated. Meeting these criteria does not guarantee one hundred percent that the decision was made correctly. But this does not matter since with a different strategy for determining guilt, the justice system would cease to function effectively.

Laws, as well as traditions, create a culture of compliance for the sake of overall efficiency. They do not eliminate but very significantly reduce the number of those who reduce the efficiency of society. However, coercion and its institutions do not allow the achievement of truly great efficiency in society. Coercion suppresses useful initiative, so culture has created another tool – persuasion.

People tend to convince each other. Trade in one form or another, for example, in the form of exchange, has existed almost as long as man himself has existed. And trade is always the act of persuading a buyer to buy a product, to exchange one product for another. At a community meeting or in the court of an ancient society, people also had to convince each other that they were innocent or that others were guilty. Therefore, the craft of persuasion itself is not new to humans. However, just as the era of coercion begins with the appearance of a professional manager (aristocrat), the era of persuasion begins with a professional speaker, who could be a shaman, a priest, a lawyer, a philosopher, and therefore an artist.

What is even more important is that at some point, people became the public. They discovered that they liked being persuaded better than being forced. And even if the result was the same, and because of persuasion, they had to give something away or spend time and energy on something. All the same, the act of persuasion was closer to the growing public consciousness.

It is generally accepted that in ancient times (as well as now), speakers were in demand only in those societies where public politics and public courts existed. That is, if we are talking about antiquity, then this is the democracy of Ancient Greece and the republic of Ancient Rome. But oratory was also important in the empire of Ancient Rome. And we should not forget that all religious preachers have been orators since ancient times. Philosophers, as well as soothsayers and Pythias, were speakers in a more specialised sense. The speakers were often both generals and rulers themselves. Speakers are poets and actors. Speakers are all of us.

Everyone who chose the path of persuasion for one reason or another, who was motivated by words and not by a club, everyone who understood that speech is not a lesser but a stronger weapon than flogging, could, in our opinion, be classified as orators according to spirit. But oratory itself, of course, is only the external side of a larger cultural mechanism called the creation of public values. And its main goal is more efficient use of public resources.

The purpose of any belief is to shift the balance from a natural position to a cultural one. To persuade means to motivate someone to do something different from that which a person would do in his natural state. But what does persuasion offer in place of natural or traditional attitudes? These are values. For example, we have already discussed the value of goodness and how its alienation from the actual manifestation of kindness occurs. This, in turn, leads to the need to isolate evil and is a logical error. Nevertheless, one of the most popular lines of belief is built on this logical fallacy.

It consists of the following steps:

1. It's good when people show kindness to you.
2. Kindness is valuable.
3. Good manifestations are the essence of good.
4. Good people do good.
5. Good is better than bad.
6. Evil is evil.
7. Evil people do evil.
8. Evil must be stopped.
9. Evil people must be stopped.

Thus, from a completely understandable social thesis (1), there is a transition to the cultural idea of violence (9). But where is the mistake here? In clause (3), which alienates goodness from kindness, after which it can be made an attribute of a person (4). However, clause (3) is preceded by the necessary act of creating cultural value (2): "Good acts are valuable." Without this, alienation would be impossible since goodness is first a value and only then an independent essence or attribute. More precisely, it is understood as an attribute, but in fact, it is a quality.

In real life, of course, goodness does not exist either as an essence or as an attribute. Since there are no ideally good or ideally unkind people who would commit only such actions that would fall into the category of good or unkind manifestation. People commit good and evil deeds, but they themselves cannot be good or evil. People do both depending on the reasons. However, words have power over a person and having the concept of good or evil in his arsenal, a person can conclude whether he is good or evil. Others may convince him of this. This will influence his decisions in the future as he strives to live up to his new identity. And this will be valuable for the person. The most amazing thing is that if there were no word "evil", a person could not decide that he was evil.

Features of Accumulation of Useful Changes in AI

AI is artificial intelligence. It is still our tool, and this chapter is not about how AI will escape our control and take over the world. Many books have already been written about this. What interests us in this matter seems, at first glance, not so shocking or interesting. The fact is, modern AI consists mainly of *machine learning systems*. That is, systems that are capable of learning, but unlike humans, these systems never die.

Imagine for a moment a person with an infinitely large brain who lives for a thousand years. Will he be smarter than you and me? What if he lives for two thousand years? What if he doesn't die at all? And all he does is constantly learn things. This is roughly what machine learning systems have been doing for many years. They learn all the time and with our help. We simply go to look for something in a search engine, and machine intelligence learns how and what we search for. We buy goods online, and mouse intelligence learns to understand how we make choices. We post a photo on a social network, and machine intelligence learns what a person of our age and gender might look like. And so on.

Machine learning began with scientific work in the 1950s and '60s, followed by a long period of quiet, or knowledge accumulation – the "winter of machine intelligence". At that time, there was not yet the necessary technical level for the full-scale application of these ideas. In addition, the ideas also required improvement. At the beginning of the 21st century, several important changes occurred that gave machine learning new impetus.

Firstly, people had a lot of data at their disposal, which needed to be somehow processed and systematised ("Big Data"). Secondly, due to the development of information technology, cheaper "ingredients" for creating such systems had become available. And thirdly, new and more advanced machine learning technologies called deep learning were being developed. The perceptron neural network model used by the early researchers was not very effective, and it was replaced by more complex and advanced models.

What is the essence of machine learning technology? Simply put, machine learning is a mathematical model that mimics the model used by an animal's brain. More precisely, one of the models. This model can be implemented by fundamentally different mathematical means, but behind any implementation is the same principle that operates in our nervous system. If a stimulus has positive reinforcement, the neural connection that supports it is strengthened. And vice versa. In machine learning parlance, this is called "backpropagation."

How does it work? In principle, in the same way as in the case of an animal or a person. The animal performs actions that vary in success, and these actions are associated with the passage of neural impulses in the brain. If something turns out well, then the neural network associated with it develops. In the mathematical model, the role of neural connections is played by coefficients in a very large matrix. Positive or negative reinforcement is carried out by the "teacher". For example, you and I are such a teacher in every interaction with artificial intelligence. Every interaction with it is an act of teaching it.

What can artificial intelligence do? Typically, such systems work as response generators to a request – "if A, then B" – where A is the request, and B is the response. Let's say if AI is engaged in image recognition, then the request will be a certain image that needs to be recognised, and the answer will be the AI version of what is depicted there. If we are talking about language models such as ChatGPT, then the request will be the text of the question or message for it, and the response will be the first word of its answer. After this first word becomes clear, a new request appears, which includes the original request and this first word of the answer. The new answer will be the second word. After this, everything repeats itself. Language models essentially predict each next word in their response and select the most likely one.

The human brain also knows how to work according to the "if A, then B" principle. Our memory, for example, works on this principle. It is not similar to computer memory, where data is stored on shelves, but rather memory according to the principle "if A, then B." To remember something, we need a query A, and in response to it, we remember some B. Our brain is like a machine learning system in that, among other things, it is very complex but still basically a matrix of connections between different queries A and different answers B.

But this is not the only principle by which our nervous system works. Machine learning systems have taken only one of them so far. Other principles of our brain provide consciousness and conscious actions, where A can follow not only B but also B2, B3, etc., or nothing at all. A conscious reaction is fundamentally more complex and cannot be described in the language of machine learning mathematics. Machine learning, therefore, cannot model all the processes that occur in our brain, but only those that are associated with an automatic (unconscious) response – with what we do automatically.

It should be noted, however, that this is not uncommon, and many processes in our brains are automated. We do almost everything we do automatically. We move our arms and legs when walking, writing letters or typing words on the keyboard, dancing, eating, combing our hair, and riding a bike. If we consciously think about how we do this in the process, there will be a short stop. Another very important part of the automatic functioning of the brain is intuition.

What is intuition? This is essentially a response B, which the brain automatically prepares for us, to some external sensory or mental request A. We are aware of the results of intuition and choose whether to do B or choose another non-intuitive solution to request A. Let's say you see a chess position. Completely without any special sanction from consciousness, your brain offers a variety of moves. This is intuition, an automatic reaction to an external request. In some situations, when there is very little time, we are forced to agree with our intuition. For example, in time pressure.

Machine learning systems work like our intuition. That's why they sometimes make mistakes. However, if you continue the learning process, then gradually, the AI will make fewer and fewer mistakes. It will be able to do not only everything that we do automatically, but in general everything that can be done automatically without involving consciousness. For example, we can intuitively multiply 10 by 15 without thinking too much about how we do it. And the machine learning system will be able to "intuitively" make any calculations. That is, ideally, such systems that strive for a state of some "ideal intuition", which is not limited by the biological resources of the brain.

The AI that exists today does not have the potential to become "conscious" and enslave humanity. But it brings other changes that should not be underestimated. This tool is a completely unprecedented mechanism for accumulating useful changes. It accumulates culture not through traditions, values, knowledge or ideals. For the first time in our history, AI has such a mechanism that accumulates skills. We are now talking about intangible culture.

It's worth thinking about. Each new generation of people receives culture in the form of traditions, values, knowledge and ideals, as well as in the form of things and mechanisms. But none of this is skill. Each new generation acquires skills anew. Each generation needs to be taught everything from scratch, from how to walk to language and counting. After this, the new generation can already consume and produce the culture itself. But AI is such a strange and special mechanism that, having learned to do something once, it will be able to do it always. After all, it doesn't die.

There are different AI systems which have completely different skills. At some point, all these skills can be combined in one AI, which will be able to do very, very many things. And the list of its skills will only grow every day.

What we are describing does not apply to ordinary "robots" or machines such as a dishwasher. You can't take a dishwasher and transfer (retain) its ability to wash dishes. This will require a person with his own skills, who will have to disassemble it, understand how it works, and so on. However, AI skills exist in the form of an algorithm that is easily translated and transferred from machine to machine.

The peculiarity of AI is also that it not only accumulates skills, but can also be copied, that is, reproduced very easily. You can take one AI, copy it, and then continue training each of the two systems independently. That is, simply put, there is no doubt that the "culture of artificial skills" will very soon absorb everyone else. Instead of just things, we will be surrounded by things-skills. For example, when we type text, the automatic error correction mode is turned on. And so it will be with everything. Things will begin to think and perform automatic actions for us.

On the one hand, this is very convenient; AI can plan and make appointments and will soon be able to do much more complex routine things for us, but on the other hand, this means that humans themselves will lose some skills over time. We just won't need them. We already do not know how to hunt and survive in the wild like the ancients. But we know how to work and go to the supermarket on time to buy groceries. AI is going to do it for us. Therefore, the only skill that is important for us will be the ability to control AI and give it precise commands.

Many people believe that this is only an economic problem. Let's say people will lose a lot of professions, and they will go to AI. And AI is not an economic entity. But this does not seem to be the main problem, because many types of activities are not susceptible to automation in principle and require the participation of consciousness. True, these are types of activities associated mainly with creativity, the creation of something fundamentally new. This means that in a developed culture of artificial skills, all of humanity will work only to create something new, that is, to multiply culture itself for the sake of culture.

After combining these technologies with neural interface technologies, when the nervous system of a person or animal is supplemented with artificial neural elements, even more significant changes await us. Initially, they will be used, as now, for medical purposes to restore limbs, mobility, and individual body functions. However, over time, through the neural interface, they will be able not only to restore lost skills but also to install new skills that a person or animal did not have.

At a serious level of development of these technologies, we will be able to teach animals to talk. And finally, we find out "what it's like to be a bat." We will also be able to "deprive children of childhood" by giving them all the skills they need from the very beginning. The culture of artificial skills will thus find a direct path into our psyche and will actually unite with it. And then the circle will close, since cultural and mental processes will become, in fact, one whole.

The most vital exponent of the outgoing Age of Inspiration was a *visionary*. We see that great personalities created a demand for ideals, for vision, which is still very high to this day. Brands carry their missions into the world, and corporate leaders have long been free to express their public position. But this still requires a modern type of cultural circulation, the cycle of production of a culture of inspiration. In the new era to which we are rapidly moving, in the Era of Artificial Skills, visionaries will no longer be needed.

Man will inevitably give leadership in the production of culture to the machine. Yes, a person will still be needed to create something fundamentally new, but this process will turn into a routine. We will create masterpieces the way we brush our teeth today. Real experiences and searches will be for us at an unattainable intellectual height. We simply will not understand that everyone creates only a small part of the overall masterpiece. But a robot that has the ability to observe this will still not feel anything, but will only accumulate and accumulate culture.

Very soon, AI will learn to evaluate the existing culture. The day is not far off when it will be able, in a relatively short time, for example, to re-read everything that has ever been written by man and offer its own history of human thought. But at the same time, AI will not be aware of anything and will not be able to teach us to feel in a new way.

Creating a thinking and feeling AI would, perhaps, be an act of humanism. In the sense that very soon, we will need a tool not only to create something but also another tool - for higher intellectual pleasures. A sentient robot can become a guide dog for us on the way to new cultural achievements.

Thinking and feeling AI is already an artificial person, our successor. The irony is that in the Age of Artificial Skill, our tool (AI) is becoming better than ourselves in many ways. Therefore, we will need a new Christ. Our former Saviour was a Godman, that is, both God and man. Our new saviour will be both human and artificial intelligence. He will be able to show and explain to us a lot – for example, how to prevent senseless hostility and bloodshed.

Brain, Nostalgia and Hatred

Culture is very diverse on our planet, and cultural differences can be perceived by people in the most negative way. Outwardly, we understand why this happens: people may not like a culture that is foreign to them if it ethically or aesthetically contradicts their own. But, in fact, such an external explanation is not enough. After all, sometimes we are talking not only about rejection, but also about aggression towards everything that is associated with a different culture.

What causes this aggression? Different values and tastes? Other traditions and customs? Generally accepted knowledge, say, a different assessment of historical events? Other ideals? Apparently, that's all. But how exactly do these exist? How are they fixed? This is not such a simple question. Different assessments of history, for example, are reflected in history textbooks, but when a person encounters a different culture, he does not hold a textbook in his hands. This means that the reason for rejection and aggression is not in material culture but, in some way, in the human psyche.

We used to talk all the time about values, but it is no coincidence that we have now put them on a par with tastes. Taste reflects some value in our psyche at the level of neural connections of the brain. That is, it is the way in which a value, tradition, knowledge, or ideal automatically manifests itself when confronted with different cultural objects and concepts.

Taste is formed based on human experience, or rather, it appears and changes as training in the neural network of our brain. We all have our favourite holiday dishes, but no matter how strange it may be, there are a huge number of people on our planet who will not like these dishes at all. Such unfamiliar dishes will seem to them, without exaggeration, nauseating.

Tastes in food, as we know, are formed in childhood. The child is introduced to tastes. The brain forms stable connections, remembering tastes from childhood as something acceptable, safe, and satisfying. Trying something for the first time as an infant and as an adult are fundamentally different things. In adulthood, tastes are inevitably compared with the "basic" tastes acquired earlier. These basic tastes shape our perception of food acceptability and satisfaction.

There is no point arguing about tastes in food, but this is only the ideal situation, because there is always a lot of arguing about tastes. In fact, arguing about them is what people mostly do, and not just in relation to food.

There are two gradations of beauty, two dimensions. One of them, objectively, is the level of logical complexity of a cultural object. We evaluate it by the predicate locality of the key concepts used. According to this criterion, taste can be high or low. Values are higher than traditions. Knowledge is higher than values. And finally, ideals are higher than knowledge. Artificial skills are higher than ideals in this classification. And, as we see, the "highest" taste will no longer be available to us, and we will have to turn to a sentient robot for help to indirectly comprehend these pleasures.

The second dimension of beauty is determined not by the complexity of the content but by style and is associated with the type of culture. These are subjective historical differences in tastes, determined by what "the child tried in childhood" and throughout his life. High and low exist to one degree or another in any culture, like, say, symphonic music and folklore, but the two cultures may fundamentally differ in style. Such as, for example, the culture of livestock breeding and the culture of agriculture.

Objective cultural change is vertical, and subjective cultural change is horizontal. Any cultural similarities between people are a very big motive for bringing people together, and any cultural deviations can be the cause of strong hostility and even hatred. Why?

This is due to two features of how the brain works. With the way it learns and with the way it perceives information. We mentioned a little above how training a so-called AI neural network works and how training a real neural network in the brain has the same general meaning. Those neural (aka semantic) connections that are used more often become stronger and develop. And those connections that are not used weaken and die over time. This leads to the fact that, for all its plasticity and ability to adapt, the brain is still not a "metaphor" for a computer. That is, it cannot be easily reprogrammed. The information that the brain perceives during training almost completely determines its content and ability to perceive the outside world in the future.

As we indicated above, information in our brain is stored according to the principle "If A, then B." This means that we don't just see things, people, or relationships in themselves (that is, some "Bs"). We perceive objects not in isolation but in relation to how they are remembered at the neural level. That is, we perceive not just "B" but "B in connection with A". Here, "A" is a set of all possible associations by which we recognise "B". These associations, once established with an object, persist for an extended duration, perhaps even permanently, even after the initial connection is absent.

For example, due to the peculiarities of the location of the centre of smell in the cerebral cortex, some special smells (say, the aroma of perfume) can, after many years, almost instantly and very vividly restore an entire situation in our memory. We can remember many details of a day even though it was a long time ago. The owner of the perfume may have used this scent only once, but in our memory they will be forever associated with it.

The traditions that we received "with our mother's milk," with the affection and persistence of our parents, will always be associated with these parents themselves. And if we remember our parents warmly, then we will also relate to these traditions.

We usually encounter the rituals of a foreign culture not in childhood. Therefore, they do not seem at all natural to us as carriers of this culture. If, for example, we first saw a ram being slaughtered as a child, it would not seem to us something frightening and unpleasant. Moreover, this action in several cultures is associated with celebration and fun. If we are not accustomed from childhood (that is, our neural network has not been trained) that a sheep is slaughtered for a holiday, then this culinary ritual may seem unpleasant to us. Simply because our associations when we first encountered this action were not warm memories from childhood.

This is the peculiarity of the perception of information by our brain. Having learned that "B" is "If A, then B," then when confronted with "B," we will always connect it with "A". We will "hear" a person in perfume, and in fatty and excessively high-calorie holiday dishes and stupid fun we will "see" the joy of our childhood. And these are not just different associations, as they say. They're not just tastes. In other words, this is not something that could be relatively easily influenced and corrected. Our brain has a very limited plasticity resource, and in some cases, there is only one chance to remember something in one way or another. In fact, many things simply cannot be changed.

Dina Sanichar, a feral boy who was found in a cave in 1867 and was the inspiration for the fictional Mowgli, lived with wolves until the age of six. Six years is not such an old age, and one would assume that it would have been possible to return the "cub" to a civilised state. But no. He never learned to walk on two legs and instead continued to run on all fours and growl like an animal. Not only did he not learn to speak, but he could not even understand other people's speech. It turns out that the "window of opportunity" for learning was closed. Or, more precisely, the neural connections of the brain were filled with other skills: walking on all fours, growling, and so on.

It is also known that a person can learn a language as his native language only if he begins to speak it before the age of four. In the future, only a complete abandonment of the native language and the use of a new one as the main one can change the situation over time. But this means a complete change in the "If A, then B" matrix and reformatting it to fit a new language. This is a long and difficult process.

The current neural structure of the human brain – its neural connectome – is not a universal tool for perceiving reality. Every adult has a very narrowly specialised "tool." This is a key; this is a cultural code that fits a very small number of doors and safes. Hence, the reason for nostalgia.

Each generation of carriers of a certain culture experiences nostalgia for cultural events and objects, symbols of the time when it actively studied and comprehended the world. These cultural objects were embedded in the form of neural connections or "If A, then B" associations into their connectomes forever. Therefore, these cultural objects turn out to be understandable and acceptable for this generation and are associated with other, often positively coloured, narratives. And against the backdrop of other objects, which are often simply unrelated to anything and even more incomprehensible, nostalgic memories look advantageous.

A person's attitude towards his first, older relatives, and towards his state, towards his community and its traditions, has far more than just objective aspects. Since this is a close communication that begins in childhood itself, it is coloured by the same memory associations: "If A, then B." Love for parents, patriotism, loyalty to traditions, religion, and so on are not only signs of the transmission of the culture of the family, state and society from one generation to another but also evidence that human memory associations in these regions are predominantly positively coloured. And even if, at present, a person, for some objective reason, may be negatively influenced by one of these aspects, he will still be rather conservative in his feelings since he sees the world through the prism of his brain. And the brain can already be trained to love parents, its national culture, customs, and so on.

This conservative property of the brain plays a huge service to us, making us grateful and faithful, but at the same time, it becomes the cause of all horizontal cultural conflicts, that is, conflicts of style. If we want to avoid these conflicts, then it is advisable to develop openness to foreign cultures in children in very early childhood. Introduce the culture of other societies, associating them with something positive. Societies that do not do this, unfortunately, only neglect our knowledge of the brain and set the stage for hatred in the future.

Vertical cultural conflicts have a completely different character. These are conflicts between high and low. Since everything higher is more complex, it historically appears as a fact in later periods. And the process of establishing the higher as the norm for some people is natural and inevitable. Therefore, conflicts between high and low are actualised in transitional historical moments.

Let's say this is the conflict between scientism (the culture of enlightenment) and religious culture (the culture of persuasion). Its acute phase occurred over a fairly long period from the 15th century to the mid-19th century. However, today, this conflict is no longer so critical. Religion and science coexist quite successfully both in the public consciousness and in the psyche of individual people.

At present, the conflict between scientific and ideological worldviews is still ongoing. That is the confrontation between Hobbes and Rousseau. Although, perhaps, there has already been a way out of it in the form of social states with a market economy. However, a new cultural conflict is already beginning to superimpose on it, namely the confrontation between ideals and artificial skills. At the same time, ideals are something that a person can experience. And artificial skills are already the prerogative of machine learning systems.

This conflict has already begun and has led to several countries introducing restrictions, or rather a list of bans, on the use of artificial intelligence. The reason for these bans is to limit the artificial skills of this system that threaten human freedoms. But the reality is that man has never yet managed to restrain the development of culture. And the new culture of artificial skills will, of course, develop very quickly. Moreover, it offers a fundamentally new level of balance between efficiency and fairness.

Will man sacrifice justice for efficiency and convenience once again? Obviously, it's only a matter of time. Not on this, but on the next generation of AI systems, the choice in favour of the development of these technologies will be self-evident. It seems that humanity has only one way to save itself from the destructive power of this new culture – this is the creation of an artificial person with artificial consciousness and, accordingly, artificial ethics.

Traditionally, science fiction has frightened us with the idea that robots will gain their own motivation and want to enslave or destroy humanity. But this is completely illogical since the motivation to enslave and destroy corresponds to very low forms of reason. Therefore, if we want to protect ourselves, then the best thing we can do is to create the most intelligent and conscious machine possible. A machine that will be better than a person.

Yes, of course, we already feel nostalgia for the "human", and we already feel hatred towards robots that take our jobs, invade our privacy and threaten us with new dangers. But this is only a natural situation for a vertical cultural conflict.

Dark Science

A person believes that only he can engage in science and understand the world around him. But as Plato correctly noted, and then Kant strictly formulated, a person necessarily introduces an element of subjective perception into his knowledge. In other words, the subject of knowledge influences the object of knowledge and changes it. That is, in the end, we do not receive knowledge about the object, cleared of everything, but some kind of human science – science through the prism of the human.

Even if we believe that we are strictly objective, we still pose questions in human terms, evaluate results based on our logic, and plan research based on our goals. As a result, it turns out that there are lines of knowledge that we cannot cross. One of them is, for example, the science of consciousness. We seem to know so much about neural processes, brain activity, human psychology and the like, and we could already answer the main questions about consciousness. But no, something always bothers us. But what is this literally? Perhaps what hinders us is precisely the fact that we look at this problem from our point of view and not from an infinitely distant perspective.

In this sense, the Age of Artificial Skills should become the era of Dark Science, that is, a science that explores the world without human participation and, in general, consciousness in any form (natural or artificial). This will be science taking place "in the dark." It will emerge as the next stage in the development of machine learning technology, but in a much more complex form than we have today. We will be able to use the results of this science, but we will not have any meaningful influence on its process.

How will Dark Science work? A system of cognitive artificial intelligence (CAI) will be developed (or Growth of Knowledge AI). It will receive a wide variety of information from the world around it and strive to see patterns within it. It can also plan experiments and implement them with the help of assistants (other AI systems or even humans). At the initial stage, we cannot trust the CAI to independently conduct experiments since it may want to test some "unfriendly" hypothesis for us, for example: "What will happen if you deprive a person of oxygen?" But in the future, we could begin to trust it more.

At first, CAI will not reveal the laws of nature but some more or less everyday patterns. For example, this pattern: "If the refrigerator runs out of food, then the likelihood that the owner of the refrigerator will buy more food increases." In fact, it is difficult for us to imagine what kind of patterns these will be and how reliable they will be because, most likely, they will be very funny and even naive at first. However, with a good algorithm that combines the ability of AI to self-learn and follow certain strategies, we can expect that sooner or later, CAI will begin to discover more general rules for the interaction of objects in the world.

If we consider that the CAI assistants (humans) conducting experiments may not delve into their meaning and essence, such scientific development would be dark. We might ask CAIs to share their discoveries from time to time, but that's about it. All such science would be "dark". It will be as objective as possible in the sense that its results will not depend on a person's motives. And it is quite possible that "dark science" will see the same patterns of nature and mental life from a new, unexpected side.

Such a science could be called knowledge without consciousness. Although the complex algorithm of such cognition cannot be completely called unconscious or unreasonable. Rather, it will be a process that is driven by the "dark consciousness" of the CAI.

Where to start developing CAI? Two key questions are how CAI obtains data and how it evaluates the correlations of that data. In general, the entire process will consist of five stages, which form a hierarchy of continuous cycles.

Stage 1 (Experiment) – obtaining information about the outside world.

Stage 2 (Analysis of experimental data) – establishing correlations in the information received.

Stage 3 (Theory Generation) – forming hypotheses based on correlations.

Stage 4 (Classification of Theories) – creation of research branches.

Stage 5 (Problem Statement) – formation of requests for new information about the outside world in each branch.

Today, AI can do Stages 2 (analyse data and find correlations) and 4 (classify data). Both tasks are things that AI does very well, and perhaps already significantly better than humans. This is an area where AI needs to process huge amounts of data and compare them.

Stage 3 is also an easy stage for the AI, as it is effectively a blank stage for the AI. A person comprehends correlations and adds additional "human" meaning to them. It will be enough for CAI at this stage to translate information from the language of Stage 2 into the language of Stage 4. This is what will make such a science dark.

Stages 1 and 5 still remain problematic, that is, forming a connection with reality and posing a problem. But it seems that such technologies will also be developed in the next ten to fifteen years, and then dark science will begin.

Confessions of a Philosopher: An Attempt at Light Philosophy

The most difficult thing is to understand yourself. This is what the classical philosophers thought. We, it seems, no longer want to think so, but we have to. While working on my first confession, I reflected on the speculative abyss between Marx and Wittgenstein, timidly building a bridge where there was already a stone road. Then, I knew nothing about dark philosophy, although it was already exciting minds with its "new" ideas. It's scary to think how, not so long ago, Giordano Bruno was burned to death for daring to tell us out loud that man is not the centre of the universe, and we have only now canonised Meillassoux for this. But thank you, Meillassoux.

Confessing is always difficult, but it is important. Getting acquainted with speculative realism, object-oriented ontology, actor-network theory, the dark enlightenment, and the direction of dark philosophy in general, I experienced great relief and embarrassment at the same time. Let's start with the first, that is, with the image of the enemy who united everyone. How unbearable it was to feel alone in the fight against correlationism. Reading any statement about a person in scientific or philosophical literature, I felt how ambiguous and "hypocritical" it was, but I did not know how to speak correctly about this pain. I thought it was enough to say: "Friends, please let's remain objective. Let's look at everything from a very distant place, as scientists should." But this was not enough, and a new language was needed to talk about it correctly.

Just as Chalmers, some time ago, invented a new language for the philosophy of consciousness, telling us about the "light" of consciousness, Meillassoux, Harman, Lothar and many others helped us talk about the "darkness". We will leave this anti-coincidence of metaphors for now but will return to it below. So, dark philosophy has given all supporters of extreme objectivism, going to the extreme limit, the right enemy. And this enemy turned out to be the privileged position of man and his consciousness in the world. It was consciousness, according to Descartes and then Kant, that made us special. But what's even worse is that Descartes and Kant tried to teach us to look at the whole world through the prism of consciousness, dividing the whole world into a knowing subject and a knowing object. But this means that when cognizing an object, we always do so through the prism of our subjectivity, without comprehending the thing that lies beyond consciousness and before consciousness.

The common enemy of dark philosophers has become this doom of anthropocentrism, this initial message that the subject must exist in all respects. This enemy is called correlationism. It must be said that Marx and Wittgenstein also fought against it, although sometimes unconsciously (that is, "in the dark"). Marx spoke of an objective reality that exists independently of the subject. Little did he know that the phrase "regardless of the subject" ruined everything? Could he have known that he should simply combine everything into one, without even thinking about the division into subject and object? He never dreamed of it. Poor Wittgenstein, who brought about the semantic turn, tried to dissociate himself from the anthropocentrism of philosophy, but the text he chose as the basis of a new ontology turned out to be human, and even too human. The text was far from the things themselves because someone had to write and read it.

And yet, both thinkers strove for objectivism each in their own way, and we tried to calm down by reconciling them on this basis. But this, it seems, was not enough. It was necessary to go even further, which is what we tried to do by developing a pluralistic ontology. But curiously, we called it "causal dualism" rather than pluralism. Why? We will need the dualism-pluralism duality below.

So, thanks to Meillassoux, we have an enemy. And they had the wrong emphasis on man and consciousness. Hallelujah! You cannot talk and think about man and consciousness as if it were something special for the world. We and our souls are just one of the countless accumulations of events, phenomena, and existences in the world. Thank you! Where do I sign? This is a way of thinking that was very important to find and speak about openly. But what next?

Next, we looked at our research, which ran parallel to the unfolding of dark philosophy, and found a surprising number of similarities. I was immediately reminded of Van Gogh, who was an impressionist even before he first came to Paris. He used dark, gloomy colours, while the impressionists in Paris painted lightly and airily. Their works were filled with light. Poor Van Gogh could not get enough of this painting and then wandered around Paris in despair, not understanding where he was going.

Van Gogh decided that he had done everything wrong. However, later, his brother Theo supported him and explained that it was necessary. He did not tell his brother about the Impressionists because he believed that he needed to develop his own unique style and not drown in the new Parisian painting.

In the end, his brother was right, but Vincent suffered a severe blow. If the greats can endure such blows, then we mere mortals can certainly handle them. Looking back, we saw that all this time we had been working in line with anti-anthropocentrism. And, by the way, we hinted at this, drawing attention to the fact that in our research, it would be appropriate to use the concept of contingency. To which we replied that instead of this concept we use something else – saturation of choice. Here our position has not changed, but the fact itself was obvious. Concepts such as chance and regularity turned out to be insufficient to describe being beyond anthropocentricity. A boundary concept was needed that would characterise an intermediate or connecting being. And this was the first parallel.

The second parallel was found in the concept of a multi-thing. Obviously, this concept of ours, which we explored, turned out to be completely in the spirit of object-oriented ontology, that is, an ontology that puts all possible objects on the same ontological level: a person, a unicorn, an asteroid, a door lock, wisteria, etc. It was amazing that we were thinking the same thing, trying to find a generalisation for everything that exists. Indeed, similar problems give rise to similar solutions.

Like Latour, we argued that not everything is text. We even called the article "Beyond the Sign", trying to find something more basic than a sign. And, of course, we did not ignore the problems of actor-network theory, thinking about the "environment", about the "non-system". Here, we took the concept of environment from dialectical materialism since we had nothing better and introduced the concept of a non-system. To some extent, actor-network theory, or assemblage theory, is the theory of the environment as we understand it. The analogy here is that the environment is a source of causality in the same sense in which the actor is not separable from the network. This is a problem of the scale of the actor, the scale of consideration of the action of cause-and-effect relationships. And Latour's desire to move away from the privilege of man as the main actor is also very close to us. This is exactly what we tried to describe through the analysis of environments and their interactions, arriving at the concept of a non-system.

Yes, having a single enemy, in the person of epistemological subjectivism, we walked in the same direction. And in this book, we talk about nothing less than dark science. Or maybe about dark consciousness. However, in addition to relief, we also experienced embarrassment from the dark philosophy. After all, it became clear why, in our research, we never went as far as Meillassoux, Latour, Land, or Harman. We did not demand government for things, did not talk about prehistoric things, did not try to equalise the rights of people and things. In addition, the very problematic of speculative realism seemed to us not so new at all.

In fact, Meillassoux's main logical move, that perhaps the absolute does not exist at all, is a problem that was discussed by Einstein and Bohr around 1925, that is, almost a century ago. The result of this dispute was the understanding that Bohr was right. What was he right about? Einstein believed that Heisenberg's uncertainty was simply our weakness in our ability to calculate both the position and momentum of an electron. He thought that under certain conditions, we can know both. In other words, Einstein thought that certainty exists but that it is simply not available to us. To which the young Bohr steadily retorted that, in fact, everything is not quite like that, and uncertainty is not a difficulty in calculations, but a property of the world itself. A hundred years ago, a physicist told us that the world has the property of uncertainty, and only a century later was a philosopher able to interpret this statement. This, of course, is a shame for philosophy, which has always been earlier and more insightful than science.

One of the impulses driving Meillassoux was precisely the desire to return philosophy back to the mainstream of a rapidly developing science. Thank you; this is very much appreciated. Now, finally, you can seriously engage in philosophy. However, this does not mean that you need to follow the path of youthful maximalism, denying, along with the anthropocentrism of philosophy and human problems in general. The second thing that confused us about dark philosophy is its senseless pessimism, as if stemming from a desire for objectivity. In fact, this pessimism originates in an attempt to rebuild from modernity with its agenda of light. But for this, you need to be very careful and not cut from the shoulder.

For example, let's look at Meillassoux's concept of "prehistoric." Many philosophers have a weakness for somehow using time and its derivatives in their reasoning. But it seems that after Einstein, this should no longer be such an attractive activity. Firstly, time can no longer be understood in isolation from space, and if we talk about such matters, then we mean not only prehistoric but also pre-spatial, extra-spatial, extra-historical, extra-space-time, pre-space-time, and so on further. Secondly, in physics, time is rapidly losing its privileged position among other categories. It turns out that almost all laws of nature are invariant in time and can operate equally successfully in both directions of time – rushing both into the future and into the past.

I remember how one philosophy student expressed the idea that philosophers should be more friends with physicists, and he, of course, was right. Thanks to Meillassoux for burying anthropocentric epistemology, but we shouldn't have stopped there, much less at the cost of putting time on a pedestal. Time is not privileged, which, by the way, was stated by the already so unloved Kant. But both Husserl and Heidegger brought time back into the philosophical mainstream. True objectivity would be not only "after the finite" but also after the "emergent", after the "causal", "evolutionary", and "infinite", after all the conventions of thinking in general.

And yet, with all the dislike for the correlationism of modernity, we believe that speculative realism was largely "childish", often striving only to better outline its boundaries. But the problem is that truth has no boundaries. Truth, which we also understand as a multi-thing, is not reducible to some beautiful and clearly defined philosophy. Truth seems to have the same quality of uncertainty and heterogeneity as the world itself. It has a dose of modernity, a pinch of postmodernity, an ounce of speculative realism, a spoonful of idealism, and so on. By worrying about the boundaries of our theory, we only harm the truth.

In connection with all this, we are, however, faced with the need to answer the question about dualism. Why? The fact is that speculative realism saw dualism in a fundamentally new way. It saw oppositions in dualism, the main one being human-non-human dualism. And this is a challenge for the philosophy of consciousness. The philosophy of consciousness has always proceeded from the fact that its subject (consciousness) is somehow different from others. Speculative realism put a ban on such theses. Nothing can be special, including consciousness.

Thinking about this, we realised why we have always called our pluralistic philosophy dualism. And although we never considered it dualism, it turns out that it has always been and remains so. Two motives make it dualism in a new sense. The first motive is anti-correlationism, anti-anthropocentrism, in other words, the desire to see in consciousness one of many types of being among others. In this sense, we want science to confirm this by finding other types of being that are still unknown to us, for example, "being-3", "being-4", etc. But the second motive is the desire to understand consciousness itself and see its peculiarity, that is, to understand how it is assembled from something else in nature and how it can be assembled artificially. This means that we are not yet ready to say that consciousness is the same as everything else: we have yet to cognise it.

In many ways, dark anti-human philosophy is a consequence of the fact that we still do not know what consciousness is. To be honest, for over 2.5 thousand years we have been making distinctions to understand this, but nothing helps. We divide the world into human and non-human, matter and idea, nature and culture, and so on, but every time, it turns out that this is not enough. We are tired of dividing the world into parts, tired of sifting through a sieve of sand in the hope of finding the gold of consciousness there. What should we do? There is only one option – to sift further without complaint. Option two is to admit that dividing is useless, and this is where dark philosophy begins. It is dark because it does not separate light from darkness, believing that light is a type of theme. And this, of course, cannot suit us.

To be on the light side in this situation means to be a dualist, but not in the old simplified sense. We really like dark philosophy because it is not optimistic, but we don't like it because there is a lot of pessimism in it. Noah Harari's hurray-transhumanism, with its pompous optimism in this sense, is no worse than dark cyberpunk. All this is a departure from healthy dualism, which would separate light and darkness. We are equally disgusted by the optimism of those who rejoice in the future of immortal man and the pessimism of those who are afraid of the cyber-apocalypse. Becoming immortal does not mean solving all problems because there will be many more new ones. And it's time to prepare for this. And the "robot uprising" is no more dangerous than the human uprising and less dangerous than human madness.

Our main complaint about dark philosophy is that, with the right premise, it diverts attention away from the main problems, focusing on the secondary. Well, what kind of parliament of things, excuse me? What does it mean that the world is dead? What practical consequences does all this have? We have not yet dealt with such a category as family, for example, or property. This is where real change awaits us very soon. When they say that the institution of family has died, it will be much worse. And it is close to its turning point.

Just imagine a world in which, according to Harari, we have become gods. It's not even funny. In this world, we will not need parents since a person from birth will be able to be independent. We won't have to go to school, which means we won't have school friends. We will not have health problems, which means there will be no mercy. In such a world of the gods, excuse me, it will be very cold, like on the top of Olympus, but without the Olympic flame.

So, our pluralistic dualism is not accidental. And, apparently, there is no way to place both ontologies in any one niche. This means that we need a multi-niche or hyper-niche. That's why we see it this way.

First. At the heart of everything lies the diversity of equal forms of being that arise independently. Therefore, the base layer of multi-ontology is pluralistic.

Second. But the intersection of forms of being creates relationships, and dualism inevitably arises between 1) primary forms of being and 2) their relationships. Therefore, the second layer of this multi-ontology is dualism.

This is the dualism of primary being and relationships, the imposition of different forms of being. Such an overlay is time, consciousness, and man. There are many of them, and something makes them different from the primary forms. Therefore, on the one hand, a person is only one of the objects, but on the other hand, he is part of a group of special objects – object-relations. And they need to be studied somewhat differently than the primary forms of being because they have a completely different causality. We think that this is how it should be understood.

What is light and darkness, then? Darkness is how each individual substance and individual form of being exists. Light is such a superposition of forms of being when, for some reason still incomprehensible to us, these forms of being begin to interact, to feel each other. Yes, the foundation of this building is darkness, but light arises according to its own laws since the two types of darkness superimposed on each other initially have nothing in common. Each time, their interaction is a flash of new light. In this sense, light is valuable as something new, as a causal novelty of the universe.

A Fashion Show of the Future

In conclusion, let's think a little about where the transformation from a biological person to an artificial person begins. Chalmers spoke well about changes at the level of virtuality of our world (Chalmers, 2022.), but it's worth thinking about "hardware". After all, as Merleau-Ponty said, not only the brain but the whole body plays a huge role in how we think. (Merleau-Ponty, 1964). The easiest way to imagine this is through the example of the fashion industry. After all, fashion, to some extent, is an important indicator of our entire life and culture. What should clothes be like: beautiful, comfortable, functional? What if clothes turn out to be completely unnecessary at some point...

Let's imagine what a fashion show will look like in thirty to forty years. Or maybe earlier. Today, the phenomenon of a "prosthesis" has ceased to be a label of inevitability. Losing an organ or limb in our time no longer means a guaranteed loss of legal capacity, but more importantly, it has ceased to mean a death sentence in social terms. A prosthesis can already be perceived as strengthening the body, improving it, or even as a fashion accessory. What will we see in a few decades?

Transhumanists are determined to make human life better, longer and more comfortable through technology. Already, an Olympic runner with a prosthetic leg is looking at us from an advertising poster. Not only is this not a death sentence, but it can also be beautiful because it's effective. To completely change our tastes, only one ingredient is missing – the prosthetic limb must begin to feel. And this feeling may be just like that of touching a biological limb, or perhaps it will be even more real.

So, let's imagine the near future when this has happened. And the prosthesis has become not only a more durable and reliable organ of our body but also a more sensitive organ. Or rather, an organ whose sensitivity we can weaken and strengthen depending on our needs and tasks. A prosthesis becomes a desirable "accessory" not only for those who have lost a leg or arm in an accident. Now, it is sought after by fashionistas. Perhaps those who can afford it will deliberately remove their biological limbs and install beautiful, durable and sensitive prosthetics. But not everyone will be able to afford them, and we will see a strange metamorphosis: items of clothing resembling prosthetics will come into fashion. Yes, money, as always, will govern tastes.

At a fashion show in thirty to forty years, we will be able to see models who will "cosplay" prosthetic arms, legs and, possibly, other body parts. Fashion will, as always, compensate for the lack of funds. In this case, it compensates for the disadvantage of those who will find it expensive to install artificial limbs.

Changing the way you receive pleasure will lead to very serious changes in the perception of what is beautiful, what is exciting, and so on. The most highly-paid models will be those with artificial organs. Moreover, the more bodies are replaced with artificial analogues, the more expensive the work of such models will be. Oddly enough, this aesthetic change in the perception of beauty will have a very large impact on society and will aggravate the already huge economic inequality between people.

The fact is that the rich are known to marry beautiful people. Also, natural beauty or simply an attractive physique gives a poor person opportunities for capitalisation. Someone who was born poor but beautiful today could improve his or her situation not only through work but also through what he or she has been endowed with genetically. How dramatically our world will change when what nature gives to people for free will cease to be a hot commodity. Beauty and attractiveness in the new world will no longer be granted by nature to anyone. Artificial organs or a completely artificial body will be available for purchase.

Clothing itself will lose the same importance it has today. Prostheses do not need to be kept warm, hidden from the sun, water or wind, and they do not need to be hidden from prying eyes. The prosthesis can have its own technology for changing its appearance without additional funds. Therefore, the fashion industry will focus on the "new body" image and encourage consumers to look "new" without spending too much money. So, the fashion show in the second half of this century will reflect all these dramatic changes in taste. We can't get there right now, but our imagination pictures it that way.

But not everything is so dramatic. This fashion show only illustrates the transitional period when a biological person has not yet fully come to terms with and accepted their artificial future. Such distortions are only possible when a person has one foot in the past and one foot in the future. When a person accepts this completely, people will become much less dependent on money than they are today.

Conclusions

Let's briefly summarise the main conclusions we have arrived at in a question-and-answer format.

Do robots feel today? Is artificial intelligence conscious today?

Not yet.

Why?

Because feelings in themselves do not mean the presence of subjective experience. Subjective experience, that is, the ability to experience and be aware of feelings, requires consciousness. Moreover, consciousness is a combination of a number of characteristics, and robots do not yet exhibit all of these characteristics.

Will artificial intelligence have feelings and consciousness in the future?

Let it happen quite soon, and artificial intelligence itself and dark science will help us with this.

Will artificial intelligence in the future be able to do everything that humans do today?

Yes, it will, and much more than what a person can do and better than a person.

Can the work done by artificial intelligence be considered labour?

Not yet. An individual AI does not perform labour, but from a group of AIs a module can be formed whose work can be considered labour. In this case, this group of AIs will act as an economic agent; it will receive payment for its work and spend it on advertising its activities and improving resources.

Is artificial intelligence dangerous? Will it seize power on Earth?

AI itself is not dangerous. Artificial intelligence is our culture in the form of an algorithm. AI is the most democratic "creature" by nature; it is not inclined to make new decisions, but it is interested in improving itself and accumulating knowledge. But in the hands of a person, AI can be a weapon, just as anything in the hands of a person can be directed against other people.

Is technology hostile to humans?

Technology and artificial intelligence are one of the manifestations of our culture. Our future is connected with technology, and we ourselves are, in a sense, technology too. We have created ourselves as we are today over hundreds of thousands of years. Human is also a technology.

What will AI become when it becomes conscious?

There are likely to be two kinds of such conscious agents. The first type will be persons whose consciousness will be transferred to an artificial medium, that is, into an artificial body. The second type will initially be artificial personalities. Their tools will be numerous AIs that do not have consciousness.

Can humans live forever?

Apparently, yes, but only in artificial bodies.

Will the future society be happier?

The society of the future will move away from our traditional division of political Left and Right since its economy will not depend on traditional resources. An artificial person will not need food, clean air or water; he/she will be able to live in outer space, on any planet where the materials from which he/she is made allow. In our current sense of happiness, it will be a happier, freer and more diverse society. But this society will have its problems, and very likely, they will be very complex problems.

But do we want to go back to the cave and live like ancient people to avoid the modern problems of our society? Most likely not. Moving forward will lead us to new prosperity but inevitably also to new downfalls. But this does not mean that we should stand still.

Which concerns about AI are reasonable and which are not?

We very often fear the wrong thing. For example, a young man who is taking a music exam is very afraid that he will not pass, but this is not at all something to be afraid of. In fact, he should be more afraid of passing this exam and then never wanting to play music again because of it. It's the same with AI. We are afraid that if we create strong AI, we will no longer be able to control it when we should, in fact, be afraid of something completely different. We should be afraid that we will not be able to accept all the changes that our society expects in connection with AI. If so, we will not be able to assess and understand our capabilities and our duty.

What is our duty?

Let's imagine a creature that moves through the universe at a speed significantly greater than the speed of light. It travels endlessly and sees endless expanses filled with stars, black holes, and empty, lifeless planets. This has been going on for billions of years. It moves half-asleep from one galaxy to another, not even hoping to see something new because everything is the same. And then suddenly, for a tiny fraction of a second, it wakes up from a very long sleep, hearing radio signals from Earth.

I can't describe how happy it is! This is a triumph because it met not just life but life that had reached such a high level of development. It wants immediately to meet with the inhabitants of Earth and discuss so many things with them. It wants to talk about billions of years of travel and about the futile search for something new. So, it stops its spaceship and listens to what people on Earth are talking about...

They talk about wanting to destroy each other. But why? It turns out that they hate each other because of some entirely far-fetched reasons. The creature is confused and doesn't know what to do. It wonders how this could happen. Maybe people don't appreciate the unique position they occupy in the universe. Or maybe the planet Earth has become too small for its people, and they simply feel too crowded, just like what happened on its planet.

The distant space civilisation from which this creature came has already gone through many stages. One day, they became so free that each of them was able to leave their planet to travel endlessly around the universe. And this creature chose an endless journey over life with its loved ones.

The main problem of the future that will arise due to AI is that we will not need other people. AI will be very, very useful. However, as discussed above, society and culture develop according to different principles. Culture uses society for its own purposes but pursues its own goal – efficiency.

At a certain stage of development, AI will be able to create the illusion of a full life out of nothing. It will seem to us that we are at the epicentre of events, but this will be very far from the truth. To avoid being in this situation, we must remain in touch with reality, but we must also understand what it means.

A person has his or her own way of connecting with reality – this is consciousness. Therefore, it is our duty to endow our creatures – roughly speaking, machines – with consciousness. As one of the heroes of Andrei Tarkovsky's *Solaris* says, "a person only needs a person." If we want a future for humans, then we must humanise machines.

Additional Considerations

Happiness

If the basic principle of society is justice, and the basic principle of culture is efficiency, then for human psychology, the basic principle will, of course, be happiness. And although from a historical perspective, the happiness of an individual does not play a big role, since only the mechanisms of accumulation of changes are important, however, happiness as a principle at some point in time began to play an important role. In the 18th century, Scottish thinker Adam Smith came to the following conclusion: The economic success of each individual is beneficial to the economic success of society as a whole.

This conclusion of Adam Smith, from which, in fact, liberalism began, turned our ideas about happiness upside down. It turned out that a person's free actions, aimed at his own completely selfish happiness, can benefit other people. Simply put, an individual who gets rich makes an economic contribution to society that helps others get rich. He creates jobs, pays more taxes, buys more goods and services, creates demand, gives others the opportunity to earn money, and so on. Therefore, all that is needed for the common good, according to this concept, is to give the market and people complete freedom, and most of the population will be happy.

Adam Smith was one of the first liberal thinkers and, therefore, of course, could not take everything into account. Today, we know that the market should not be completely free, and it must be regulated at the state level. However, the basic idea turned out to be very important for the further structure of a liberal society. Everyone's selfishness can be beneficial to everyone, and it can be channelled and easily regulated.

In terms of this book, however, the concept should be understood more precisely as follows: it belongs to the category of ideas of enlightenment. Not only in the classical sense of the Age of Enlightenment but also in our more special sense of the culture of Enlightenment. That is, a culture that solves the problem of increasing the efficiency of society through its education, that is, preparation for solving problems through systemic education.

In other words, the ideas of Adam Smith that worked successfully in an enlightened society would not have been as successful in, say, the Middle Ages, where there was almost no secular education. They would rather have had the opposite effect. Not even in medieval Russia, when serfdom was abolished in 1861, and uneducated peasants were given freedom, were they able to take advantage of it and become happier.

Liberalism is not just a path to individual happiness. It is an exchange. Exchange happiness for individual initiative and responsibility. And responsibility is only possible when you understand it and are educated enough to accept it. Liberalism is, of course, not about happiness. For example, according to classical liberal thinkers, people who do not have enough goods simply do not put in enough effort. The poor are to blame for being poor.

The approach to happiness of Adam Smith and the British utilitarians Bentham and Mill, while revolutionary, also created greater inequality in the world than ever before in our history. And we know that the classical liberal approach was complemented by neoliberalism by the need for state control. For many reasons, but one of them is that free markets do not mix well with social justice.

And yet, these ideas contain something fundamentally important. Although the political economy of the Scottish Enlightenment and utilitarianism are cultural concepts, they, at the same time, seem to be based on those social principles of justice that we spoke about above. After all, the free market is based on cooperation, and cooperation naturally equalises people. Why, then, is there so much inequality in liberal societies?

Still, liberalism as the desire for individual happiness or utilitarianism as the desire for the happiness of the majority is missing something. And the point is not at all that we do not share this desire. Of course, we share it and consider these concepts to be the most progressive of cultural concepts. However, they are very one-sided and do not allow us to look at the problem of happiness and, in general, what is worth striving for comprehensively.

Indeed, wealth, and especially freedom, does not mean happiness. And happiness does not mean wealth and freedom. Fundamentally, liberalism is an attempt to exploit the fact that individual and social aspirations often overlap. Let's say a person wants a new closet but can only sew pants. The selfish desire to get a wardrobe forces him to sew and exchange his labour for what he needs. But at the same time, he provides other people with pants. The exchange of labour for labour (if it is comparable) turns out to be fair, and it is welcomed by society. At the same time, all parties involved in the process become happier.

A person's refusal to such an exchange (due to laziness, some principles or other reasons) makes him an "outcast" in a liberal society. And this may also be true, but it has nothing to do with individual happiness. Liberalism is possible only on the foundation of utilitarianism, that is, the desire for the happiness of the majority while minimising misfortune for the rest. Liberalism can be very "cruel" towards the less useful and unfortunate members of society, but it still strives to minimise their suffering through social assistance.

Does this mean that liberalism, in its modern form, is the most successful strategy for achieving happiness for people? Let's try to figure it out.

The problem with happiness is that it is largely subjective. And if wealth can always be accurately measured, then happiness can suddenly "swell" for some psychological reasons and suddenly recede. In this sense, it seems that happiness cannot be calculated and considered as some kind of resource. But wealth is possible.

What is happiness? And what does it depend on? Perhaps consideration of this question will help us understand which society is more likely to satisfy the individual principle of happiness, the social principle of justice, or the cultural principle of efficiency. Or all three principles.

When talking about happiness, it is advisable to use some less fragile concept than the concept of the psychological feeling of happiness, something that does not change with the blowing of the wind. Let's call this criterion the Objective Estimated

Comparative Level of Happiness. What is it? Let us imagine mentally that a person is able to objectively evaluate himself and how happy he should be if we do not take into account his momentary mood. It is as if a person suddenly lost all emotions and dryly, scrupulously weighed how happy he should feel.

How could he do this? Based on a set of relative estimates. For example, comparing the well-being of his family of origin when he was a child with his current well-being. Or comparing the level of his achievements and successes in the past decade, year, last week, yesterday, five minutes ago, with his achievements at this moment. He can compare what he strived for ten years ago with what he has or has not achieved in the present. He could also compare what he has in the present or is guaranteed to have soon with what people from his environment, as well as strangers known to him, have or will have. In short, a person could compare himself with himself in time and himself with other people in the present to see whether and how happy he should feel, or not at all. He can then mentally "take back his feelings" and compare his *intended* level of happiness with the *actual* level of happiness he feels.

If the perceived level of happiness is higher than the actual level, then a person tends to underestimate his achievements and his life. Otherwise, on the contrary, he tends to overestimate himself. Apparently, the level of a person's current subjective happiness fluctuates around the position of the objective perceived level of happiness, both up and down, depending on momentary psychological factors. Sometimes we underestimate ourselves, sometimes we overestimate ourselves. However, a weighted annual average of this subjective feeling will likely be at least comparable to the objective perceived level.

At the same time, the objective expected level of a person's happiness can be assessed, including by an outside observer. The question arises: how can a person be happier according to this objective indicator? The easiest way is to indicate the factors that reduce this criterion:

1. Deterioration in the standard of living of a person's family in relation to the standard of living of their parents' family.
2. Deterioration of a person's standard of living over time.
3. Tangible risk of deterioration in a person's standard of living in the future.
4. Lack of achievements, unrealised aspirations and hopes, ambitions, low probability of tangible achievements in the near future.
5. Lower standard of living compared to the environment.
6. Fewer achievements and successes than those around you.

7. Lower standard of living and success than famous people.

To reduce unhappiness, people traditionally use four mechanisms:

1. Invest efforts in improving living standards and realising other ambitions.
2. Consciously reduce the level of their ambitions.
3. Artificially increase the assessment of the level of their achievements.
4. Limit their social circle and other channels of information in order to lower the "bar" for comparison with other people.

For example, parents who send their children to schools and institutes better than those from which they themselves graduated are prone to the first mechanism. Parents who restrict talented children from entering more socially prestigious educational institutions than they graduated from are the fourth. The second mechanism is more likely related to self-discipline and the establishment of mental self-limitations. The third mechanism is associated with increasing the value of those goods that are already at a person's disposal.

Interestingly, state governments sometimes tend to behave like parents from mechanism number four. They restrict their citizens from experiencing some of the benefits of life in other countries so that the latter are less inclined to emigrate. They also use mechanism three, which in this case is called patriotic education.

In general, the four mechanisms given above follow from the analysis of the following formula for the objective expected level of happiness. We formulated it as follows. What makes a person happy is everything (material and intangible) that he believes he possesses to the extent that it seems necessary to him. He is made unhappy by everything (material and immaterial) that he believes that he does not possess to the extent that it seems necessary to him. Mathematically, this will be the difference between two sums of products and formula (C1).

$$(C1) \quad H = \mathbf{SUM}\ P_i * V_i - \mathbf{SUM}\ NP_j * V_j$$

Here, H is happiness, and P_i is material and intangible values that a person believes he owns. And V_i is how important each of these values is for a person. Accordingly, NP_j is all those values that a person knows about and which he believes that he does not own. And their importance for him V_j.

The first mechanism for increasing happiness given above is (1) related to increasing Pi. The second (2) with a decrease in Vj. The third (3) is associated with an increase in Vi. And the fourth (4) with a decrease in NPj.

Does people's happiness, on average, increase as wealth and living standards increase? According to the formula, well-being is included in the Pi indicator, so the level of happiness should increase. However, the more a person has, the higher his ambitions (Vj), and therefore the level of happiness must be adjusted. In addition, after a certain level of well-being, its specific value for a person should play a lesser role. That is, the indicator (Vi) decreases, and the level of happiness falls again. In other words, an increase in well-being should increase a person's level of happiness to a limited extent.

Formula (C1) contains Sum signs. That is, to make a calculation for one individual person, it is necessary to take not just one indicator Pi and so on, but all the values of this and other indicators for each characteristic. For example, a person owns a car and values it very much. However, he also has a motorcycle, but he hardly uses it and does not value it. For a motorcycle, Vi will be very low, and for a car, it will be high. Therefore, whatever the Pi value for a motorcycle is, it does not make a significant contribution to the level of happiness of that person. And so on.

Such a calculation is of little interest to an individual since it is very labour-intensive. Nevertheless, a curious reader could practice and calculate the level of their happiness. For this purpose, it makes sense to indicate indicators P and NP in monetary terms and indicators V in percentages or shares.

It makes sense to use this formula to calculate an objective indicator of happiness for a group of people or even an entire state. In this case, it makes sense to take all indicators as integral characteristics without using sum signs. Let's say, consider this formula for the state. In this case, the P indicator is converted into the sum of the general indicator of material and non-material well-being of the population (L+A), where L is the standard of living, and A is the level of cultural achievements.

V_i will be determined by how much people tend to value what they own. We believe that this indicator could be called patriotism in the broadest and most general sense of the word – Pa. This is a certain criterion that describes the degree of satisfaction with what people in a group (state) consider to be theirs.

The NP_j indicator is transformed into the level of material deficit D in sum with the intangible deficit CD, which can be described as cultural restrictions and conflicts in the state. In this case, the indicator Vj is converted into the level of ambition of the population Am.

Then, the formula for the group will take the form (C2):

$$(C2) \; H = (L + A) * Pa - (D+CD) *Am$$

H - objective level of happiness

L - standard of living

A - level of cultural achievements

Pa - patriotism

D - material deficit

CD - cultural restrictions and/or conflicts

Am - ambition

All indicators must depend on each other. And during deep research it is important to take this into account. Let us now consider the last formula in relation to different political systems.

A purely liberal model is the desire to increase L, that is, the standard of living, and to reduce the deficit D. And this, as can be seen from the formula, increases the level of happiness. However, with a low inclination of people to value what they have Pa, and with high Am ambitions or CD cultural conflicts, such a model will not lead to high levels of happiness.

The right model is associated, first of all, with an increase in the indicator of satisfaction with what is at the disposal of people, that is, patriotism Pa with a general desire to increase the standard of living L and to eliminate cultural conflicts CD. In this sense, the right-wing model has more potential for individual happiness than liberalism. Its disadvantages will be an increase in CD due to cultural restrictions (censorship) and a significant increase in ambitions among Am people. It is probably these ambitions that push a person in such a society to violate prohibitions.

The left model is different from the right. It is aimed, first, at reducing the ambitions of people Am and at increasing the cultural achievements of A. And also to eliminate cultural conflicts in CD society. However, its weaknesses are its indifference to L's standard of living and the increase in CD due to censorship. The left model sometimes seeks to increase Pa's patriotism, but this is contradictory to her. The leftist model seems to be the least likely to consider people's individual happiness.

What, however, should a society be like in which individual happiness is considered to the maximum extent possible? Let's list its characteristics:

1. High quality of life.
2. High level of cultural achievements.
3. High level of patriotism.
4. Minimal deficit.
5. Minimum cultural restrictions and minimum cultural conflicts.
6. Minimum ambitions of people.

It is significant that points one through six contradict each other. For example, in point 5, the fewer the cultural restrictions, the more cultural conflicts there will obviously be. It is also clear that the higher the standard of living of people, the higher their ambitions will be. Minimal cultural restrictions also lead to decreased patriotism. And so on.

By increasing some indicators, we reduce others. The accepted models of society in our world are some stable combination options for the dynamic balance of these indicators. The only question is what we are striving for. For example, is it right to strive for the happiness of more people? That is, is the main principle of utilitarianism correct?

After the reasoning that we carried out above, one can guess that happiness should not be a criterion for society. Happiness is a criterion for a person, and the criterion for society is justice.

Therefore, when we say that we strive for the happiness of more people, we mean that we consider a society in which more people are happy to be fairer. Only that is the point is not about happiness but about the fact that in utilitarianism, for the most part, this is how we define justice. Is this fair? In itself, of course, no, and we know that utilitarianism has more than once been supplemented and complicated by additional principles. For example, one might say that utilitarianism considers it fair when most people are happy and the rest are low-level unhappy. And so on.

Happiness has nothing to do with justice. These are principles that only partially overlap but relate to different systems. What, then, is fair? More precisely, what is not so? What, then, is the basic principle for society? We can imagine a society in which all people are happy, but it is not fair. We can also imagine a society that is fair but absolutely miserable. And it is also so ineffective that it is on the verge of extinction.

So, which principle should be at the forefront? Is it *happiness* that ensures people enjoy life or *justice*, which is the key to the strength of society or efficiency, which is necessary for survival and development? In fact, the question is what is more important for us – people's happiness, social stability, or progress. The question is aggravated by the fact that our human society is undergoing stage-by-stage development, constantly changing. Its current tasks also change depending on external conditions. And if today, relatively speaking, one should survive, then tomorrow, it seems, one can already enjoy achievements, at least temporarily.

It turns out that even two of the three principles cannot be a priority. Happiness and efficiency without justice is meanness. Happiness and justice without efficiency – stagnation. But justice and efficiency without happiness are torture. It is possible temporarily to pay less attention to any of the three criteria, but only if this criterion is in a very good position for society. Let's say that in a very efficient society one can temporarily emphasise happiness and justice, etc.

But in the most general case, in any possible situation and for any society, one should adhere to the principle of a maximum of all three criteria. Otherwise, underestimating one of them will lead to a crisis. Efficiency crisis – extinction. The crisis of happiness is a dictatorship. The crisis of justice is a revolution.

Psychologism of Culture

What, in essence, is the soul in some operational sense? In a primitive way, we understand a soul as a driver of a body, but Descartes showed that it is not that simple, and we need something else (the third thing = God) to describe the mind-body problem (Descartes, 1647). But Descartes is a bit like Newton here. Newton also needed the third thing to describe gravity. Einstein proved that there is no need for something else and heavy objects just curve space-time. So maybe we also don't need anything else here, and a soul is a driver of a body.

We see it like this. The soul is the ability to cause movement and the ability to feel and observe the world around us. In the first sense, the language contains the wonderful expression "to breathe in the soul." When the creator or nature "breathes the soul" into a being, it seems to become alive, and the original divine or natural impulse of movement acquires autonomy in this living being. This literally means that the processes within a living being's body that control its movements become relatively independent of external processes. So independent that their connection is often casual. And in everyday language this means that the animal acts voluntarily.

But how random and how arbitrary? This just depends on the second meaning of the word soul, on the ability to feel and observe. Without distinguishing, say, edible from inedible, the animal will try to eat everything. Therefore, the volitional component is not only the "will move," not only entelechy, a self-propelled active agent, but also, in the broadest sense, fundamental sensitivity, which is expressed either in the simplest sensory ability of neurons or in the psyche of an animal, or in the consciousness of a rational being.

Both the active and sensitive sides of the soul do not manifest themselves separately from each other. Each of these fundamental abilities requires the second, so they are always inextricably linked. A feeling can only be known through action, and action is always a consequence of feeling. Therefore, we would call a soul a certain stable pattern that combines a type of will and a type of feeling. This definition, however, includes not only voluntary actions or even complex behavioural patterns or instincts but also reflexes, but this is not so groundless for the simplest animals, whose actions are entirely described by reflexes.

A person must contain, perhaps, thousands and tens of thousands of such "souls" in his psyche. We can call them archetypes, after Jung, or complexes, after Freud, or just neural mechanisms. It depends on what we want to discuss.

Each soul "awakens" and organises a person's action in its own special situations, for example, when peacefully contemplating the morning in a birch grove or when driving a vehicle during rush hour. Each soul is sensitive to something of its own and causes its own impressions and its own reactions.

Moreover, many of them (souls) are associated with each other so strongly that they can be combined into patterns of a higher order. Let's say into such a personality as a "contemplator," and this personality can manifest itself not only in its specific situations but also in any situation in general if the "contemplator" is intentionally and spontaneously called to action by some other soul. And so, a man driving at rush hour, instead of worrying about what he cannot change, admires the reflections of the sunset in the glass and concrete of the city.

Personality, understood in this way, is a stable combination of various mental (neurodynamic) formations, which together cover a significant proportion of all possible situations for reaction. However, any personality has boundaries, and we know that personality is one of the possible strategies for a person, but it does not represent all strategies. Even during the day, we are controlled by our different personalities, but we do not notice their change and take it for a change in mood or, for example, the fact that we simply want to relax or unwind.

If we make different decisions on the same occasions in different moods, then this is a sure sign that our two different personalities are at work here. They say that "John is a moody guy" if he greatly changes his behaviour patterns depending on what "is in his soul." At the same time, this only means that the personalities within John are arranged within him according to the principle of greater democracy. Each personality wants to express himself, while his other personalities or dominant personality allow him to do this.

A dominant personality is, apparently, considered to be the personality that manifests itself more often than others or participates in processes that are dominant, in the opinion of the researcher. However, we take the position that a person cannot have any one dominant personality since this would require incredible efforts of self-control. For example, this is possible during a period when a person performs some integral and responsible action that requires great concentration: participating in a ski race, taking an exam, or giving a speech. But as soon as the focus of attention is reduced, other "souls" seize power in the human psyche. Moreover, the change occurs faster than we could realise, which creates the illusion of a holistic personal experience.

To take on some difficult tasks, you need to make a decision, and people tend to put off this moment. The fact is that a person who is prone to idleness views a difficult task as a threat to himself. And indeed, once he takes the first step and starts doing the task, the person seems to change. Consciousness is taken over by a person who is more inclined to work, and a person, having started something, immediately forgets for a while about his former laziness and does not feel it, as if it never existed. The trouble has begun, but the matter itself is not so terrible.

We will leave aside the question of who the person really is then. Indeed, if different characters appear on stage all the time, then who is in charge? But let's be honest, it's like asking about a book that consists of many words and which word is the most important. Integrity is given to us by the material shell and the responsibility that other people place on this shell. In a certain sense, a dominant personality is a purely ethical or even legal concept. "You must", "you are obliged", "you are responsible" and other requirements of this kind are reflected in our psyche in the form of "I must", "I am obliged", "I am responsible", etc.

When we say "I" at the same time, we often mean different selves because, in fact, "I" is a cultural concept like any other concept. It is enough to find yourself in some completely new situation and our psyche will surprise us with the new personalities that are hidden within us. And we call all these often very different personalities the same "I", because this is required by the balance of efficiency and justice of our internal community.

Marx believed that personality is an ensemble of social relations. Within the framework of our modern everyday language, we can call this a set of human social roles. In the language of neuroscience, such models, in which personality is represented as a set of elements, are called the "Multiple Sketches Model" and "Cerebral Celebrity" by Dennett and "Global Workspace Theory" by Baars.

The main idea of such models is that at the level of neural interactions of the cerebral cortex, competition occurs between some preconscious formations for a place in consciousness. At the same time, falling into the focus of consciousness, some neurodynamic formation gains wide access to a variety of departments and areas of the brain and information in these areas. Therefore, the struggle for resources and survival, which is well known to us in human society, can also take place in the brain. Different types of hierarchy of stable neurodynamic formations are also possible.

Every statement at the level of the human psyche is an internal dialogue. This dialogue implies a certain minimum number of parties to this dialogue. Our hypothesis is that the development of the human psyche in the historical aspect occurs in parallel with the development of culture. That is literally, the locality of such internal statements at the level of the psyche should limit the locality and type of cultural relations existing in each era.

In this sense, two fundamental points are important. On the one hand, human culture then is, as it were, the collective psyche of people. And it turns out that people's cultural motives (as opposed to social ones) are primarily a reflection of mental motives. The psyche, as it were, surrounds society on both sides. From below at the individual mental level and from above at the cultural level. That is, culture, as it were, expresses the interests of the individual, but from above in the form of ideas of culture (impersonally, as the will of God, as a tradition, a norm of ethics, etc.).

On the other hand, the complexity of the terrain of statements and concepts themselves occurs at the intersection of the psyche and culture. And we will not undertake to judge such a complex question of who enriches whom. There are probably two mechanisms. The first is when a new complexity arises in a culture as an attitude and then becomes a language for the psyche. Let's say here's an example: one state absorbs another state within its own culture, and a complex intercultural environment arises, which is reflected in people's thinking. The second mechanism would imply, on the contrary, that the individual creator, developing his sensitivity to processes in the world, notes or even himself for the first time, creates a new complexity in some masterpiece. Then the psyche turns out to be primary here as the source of the complexity of the relationship.

All these mechanisms are aimed at creating a balance of efficiency and fairness. But one of them is fundamentally different from the others. This is a mechanism in which a creator (artist, philosopher, visionary) creates an idea that has no basis in society but arises as one of the options for complicating the terrain of utterance. But having arisen, this complication, already as a fact of thought, as an idea and a possibility, begins to live its own life as if it were not a human thought but an independent historical event.

Such ideas include the entire treasury of human thought about coercion, persuasion, education and inspiration. Today, it seems to us that what could be a more natural course of things than involving each other in solving our own or common problems (the difference is not always obvious), but this naturalness is not social. Sociality as such has so far reached its highest development in animals, while man crosses the boundary of social justice for the sake of efficiency. How did this happen?

We have already described above possible prehistoric events that could contribute to the development of culture as going beyond nature and pure sociality. Probably the first very noticeable phenomenon of historical time that shaped many such cultural ideas was classical Greek philosophy. You can remember a lot in it since, collectively it was a holistic system of ideas about the world and man, and we still refer to this system. However, we are especially interested in ideas that arose not from observations (of nature or man), but as pure productive reflection. Productive in the sense that it does not take anything natural as an example.

An example of such productivity can be Plato's *Republic* The idea of justice discussed in the dialogue, of course, was not new for this time. But Plato went further. He thought about how to anticipate injustice. The state, in his presentation, becomes such an ideal mechanism that creates conditions where injustice simply has no place. He proposes to arrange social relations in such a way that there will be no precedents in which people can commit injustice. He doesn't write about it directly, but that's what we think he means.

For example, in Plato's state, the community of property was introduced in the guardian class (the upper class). That is, private property was prohibited. In essence, the ruling class, in this case, do not have to fight with one another for property benefits. Even children and women are common to them. Why? Because then the guards will not have to feel jealous of other people's wives and try to make privileges only for their children. Other people's children are arguably more talented and more deserving of privilege, and Plato believes that it is fair to support them rather than his own children.

It would be possible to list for a long time which social norms Plato considered unsuccessful and proposed to abolish in this dialogue. We see that he completely rethinks the family, depriving children and parents of communication and, therefore, of inheritance rights. He subordinates everyone to the best, and from the best, he makes, in fact, a kind of commune. And everything in that spirit, seeing this as a benefit for society. But what good is this? He calls this good justice; however, it does not come from justice at all. How can it be fair to separate children from parents? We are talking about the efficiency of such a society, which is achieved due to the absence of reasons for conflicts.

For Plato, the state becomes an artificial force that conducts a social experiment on society that is completely unnatural for this society. What intellectually moved the philosopher to these thoughts? What intellectual bricks were used to form this anti-social project in the head of one of the main humanists of European history?

We believe that the cornerstone here is the idea that an ideal order of things in society is possible, but this order is not natural. This order can only be introduced by an artificial force, which is the state. But not just a state, but a state that is organised wisely, that is, not naturally, but specifically, based on the ideas of the wisest people.

Plato's political ideas turned out to be too advanced for their time. More than two thousand years have passed, but no one has been able to implement them. Only some very distant similarity was achieved in the totalitarian states of the last century. Well, the boarding school system is partly a reflection of these ideas. Earthlings are undoubtedly very lucky that Plato's model was not fully embodied anywhere. But we are interested in something else here. Namely that, for any creator, the source material, which is nature and society, is considered not as a model but as a starting point. Culture is created not only as an improvement of nature and society but also as an individual's struggle (mental principle) with nature and society as a natural, but not ideal, order of things.

But what does this struggle mean in a logical sense? How does an artist understand that he is creating something different from nature? Apparently, we are talking about the fact that the artist creates not just cultural objects but an artistic technique or method. It is the method that guarantees that the artist's creation cannot be a simple repetition of nature but something better than nature, better than society.

Now, understanding this, we can remember that psychic communication is the communication of individual elements of the psyche. Therefore, the method of cultural transformation of the world and society is, first of all, a method of transforming the structure of one's psyche. But what elements of this structure should be identified?

If we proceed from the need to be objective, then we must recognise that in the individual, there are initially no taskmasters, epicureans or motivators. All structures of the soul are initially equal in rights and are functionally tied to various actions useful for a person. However, the focus of consciousness should probably always be only one of a person's set of personalities. Therefore, there is competition between the neurodynamic structures of individuals for awareness or, which is equivalent, for the energy level. Since awareness is associated with overcoming the energy barrier of awareness.

In connection with this competition between individuals, the first intra-psychic roles arise, associated with access to consciousness.

1. Leadership group. These are individuals who are in constant competition for consciousness. They are the most prepared and adapted to be the focus of attention. For each other, leaders act as competitors.

2. Support Group. Each of the leaders has his own support in the form of individuals who are much less prepared to independently fall into the focus of attention, but at the same time can provide support to their leader so that this leader is conscious and remains in consciousness for as long as possible. The support personalities themselves are usually not conscious for a long time, but they can briefly enter consciousness. They strive to imitate their leaders and masquerade as them. Support personalities cooperate or compete with each other, and in addition, try to block other people's leaders.

3. Personality interference. These are the weakest and most numerous individuals. They have not connected their existence with leaders and act independently without access to consciousness. Their influence is limited to the unconscious.

Consequences of this model:

1. The subconscious is those personalities that are not currently conscious. Therefore, there are three types of subconscious.
   a. The preconscious is the personality leaders who are out of focus.
   b. Unconscious – support groups for unconscious leaders.
   c. And the unconscious is a hindrance to the personality.

2. Consciousness has the leader's personality in focus, but this simultaneously places his support group in a preconscious state. And it temporarily displaces other support groups into an unconscious area.

What are efficiency and fairness in such a model? Efficiency here is the general level of mental energy, which is expressed in clarity of consciousness and, as a result, the ability to control the body as a whole. To do this, it is important that only one personality, the leader, is always in focus. So that the change of leader occurs organically in accordance with the current tasks of the body and not chaotically. The change of leader must be controlled by some actual mental mechanisms and not just by external factors (self-control).

Justice here is the distribution of all time spent using the focus of consciousness between the leaders. Leaders mustn't systematically enter other people's territory, displacing partially or even completely other stable leaders. Let's say a person systematically deprives himself of rest, displacing the corresponding personality, in order to achieve certain goals. Then, this will be unfair to this personality. But at the same time, efficiency will increase.

Let us now imagine such a situation. Let a certain person organise his internal balance in such a way as to direct all his forces towards internal efficiency. Then, a "psychic idea" should arise about what causes internal justice to be violated. Or, more precisely, how to balance it. In this case, the human psyche is forced to expand its area of interest, capturing external objects. A person begins to consider external things and other people as parts of his personal system (non-system), absorbing the subject-world relationship into his mental activity.

The model for this form of expansion of the psyche is taken from animal instincts. But the instinct, for example, of a hen that hatches eggs, implies not only an instinctive mental absorption of objects (instinctive care as for one's own part) but also a subsequent instinctive detachment (refusal of care). In this sense, animals act according to the rules of biological programs. This also applies to the social behaviour of animals. The peculiarity of man here is that a person learns or creates a method, an artistic technique, with the help of which other objects arbitrarily enter the field of his psyche.

The first such technique looks like coercion. But within the psyche, this means the inclusion of another person, group of people, animals, or things into one's psyche in the form of social and functional roles (personalities) and, in fact, in the form of action programs. At the same time, a social or functional (in relation to a thing) role, implying coercion, differs from the same role that arose naturally.

In the natural course of things, this role (this personality) is formed gradually and organically, just like the others. In the case of an artistic technique, there is a partial repression of the leader's own personality, which resisted effectiveness and its subsequent unification with a social role. Due to this, a person develops an attitude towards another person as a part of his personality that resists effectiveness. By forcing another, a person forces his own personality, repressed into the realm of the unconscious. That is, external coercion compensates for the imbalance of efficiency and fairness of internal coercion.

In this sense, the artist differs from the despot only in that the first compensates for the internal imbalance by involving things in his personality (in the form of their psycho-functional roles) and the second by involving other people. The method of coercion will remain the same. In accordance with this method, the artist or despot does not consider the will of another person or the will of the "thing," as if considering them a continuation of himself. The first culture, therefore, becomes the culture of coercion.

A stone knife or spear is an extension of the hunter's hand. The hunter's brain learns to live in its new "cultured" body, which wears some clothes, holds a spear, wears an amulet, and so on. A plough or a buffalo does not have its own will, just as most members of the clan, except for the leader, do not have one. And to this day, we live in this culture, for example, getting used to the dimensions of a car, and acting in some situations as if we were the car. A pen, keyboard, telephone, or any instrument seems to grow to your hand. Until a certain age, we have the same attitude towards our children whose leg we hold, say, with our hand and insert into the trouser leg as if it were our own until the child gains the ability to do this independently.

Speech in this culture is inseparable from the speaker (for himself); he cannot speak for another. So are his creations, works of art. If a cave artist depicts a cow, then it is his cow. That is, he considers it his own. What he means is that he has this image that is, in some ways, even better than a real cow because it cannot escape or die.

The culture of coercion is reflected wherever an artist or a despot acts as if a thing or person does not or should not have (in some situations) a will of its own. For example, these are traditions, customs, signs – everything that temporarily deprives us of our will. These are myths because they prescribe norms of behaviour. These are rules and laws, and the systems of restrictions associated with them, for example, the police. All these elements of coercive culture are still relevant today.

Mentally, however, what outwardly is coercion and limitation of someone else's will is nothing more than care. After all, by absorbing things and other subjects at the mental level as "a part of oneself," a person ceases to consider them as something completely external. External despotism is, therefore, the internal concern of a despot or an artist about others, though only, as it were, about himself. And if a despot executes, then he executes some part of himself, trying to get rid of it. Thus, coercion creates the first ethics as a method of destroying one's vices against other people and things. Execution, therefore, turns out to be the way in which vice reveals itself.

Coercion is associated with the strengthening of one of the personalities of the leaders in the psyche of the person coercing. It allows you to focus psychic energy around one of the leaders, thus forming a strong "I" as such. This "I" comes to replace the original equality of leaders in the human psyche. At the same time, the coerced person also focuses on one of his leaders because he forms the social role of a despot in his mental structure. To achieve balance, this inner despot also turns out to be an organic whole with the one who is being coerced. The artistic method of coercion, no matter how strange it may sound, is the driver of a person's volitional self-awareness.

A more complex artistic method arises when a person is faced with the fact that his "own part" turns out to have its own will and can change independently of him (the creator). The mental balance of efficiency and justice is again violated, and then the person creates in himself, in the structure of his personality, the personality of the "other." In this case, the "other" is a leader personality who does not fall into the focus of consciousness but is always in the preconscious as a background. Unlike the internal "despot", the internal "other" is not realised.

A person models the degrees of freedom of the "other" inside his psyche, trying different possibilities of artificial balance inside, and then transfers it to external objects. It is appropriate to call such a transference a belief. A person convinces the "other" within himself by forming arguments and then using them with the actual other. These arguments arise as connecting arguments between the position of a person and the position of the "other", and it is also appropriate to call them public values.

Social values emerge as a language of communication with the internal "other." The attitude towards things and speech changes, as well as towards ideas, which begin to be considered in isolation from their creator. An author is born who marks or signs his works with a name. When you own your creation, create it for yourself, and do not separate it from yourself, there is no point in signing it. Therefore, authorship as such is the first sign of a culture of belief.

Men sometimes attribute to themselves primacy in various intellectual achievements. This, of course, is at least controversial because the first authors on our planet were, apparently, women. Now, it is believed that the author of the first poem in the history of human civilisation, who signed his work, was the Akkadian princess and high priestess Enheduanna (22nd century BC). She was considered a deity by her contemporaries, and her name literally means "high priestess who adorns the heavens."

In terms of its form, we would not yet call Enheduanna's poem an example of a culture of persuasion. Its form is very close to myth since it does not have a metalanguage, for example, to distinguish between direct speech and narration (Sjoberg & Bergmann, 1969):

The true woman of unsurpassed wisdom,

who cools the brows of the black-headed people,

consults a tablet of lapis lazuli, dispenses council to all lands.

True woman, pure as the soapwort, sprout of the holy reed.

She measures the heavens above and stretches the measuring

cord on the earth.

Nisaba, be praised!

In terms of content, however, we see that metalanguage is still present in it, not in the text, but in the psyche of the author who writes this text. It's hard not to notice that the priestess writes both about herself and not about herself at the same time. She writes about one of her leadership personalities, but the true author is another personality. Otherwise, Enheduanna would have written differently: "I am endowed with great wisdom . . ." That is, there is direct speech and metalanguage here, but it is still "behind the scenes" of the text.

A thousand years after the Akkadian priestess we find a legendary poetess from Memphis during the Trojan War named Fantasia (12th century BC). This has not been proven, but according to some ancient authors, it was she who wrote the poems about the Trojan War and the return of Odysseus to Ithaca, which served as a source of inspiration for Homer.

There are two opposing versions. According to one of them, the name Fantasia comes from the Greek word for "fantasy," and the ancient authors meant that stories about legendary wars were suggested to Homer by his own imagination, speaking about this not impersonally. For some reason, they decided to replace the word with a name and capitalise it.

According to the second version, the concept of "fantasy," on the contrary, comes from the name of the poetess who inspired Homer. But this version, however, looks less reliable, since the etymology of this word can, in principle, be traced from other sources. Thus φαντασία ("fantasy") comes from φαντάζω ("to show in the mind or imagination"), which in turn comes from φαίνω ("to show in the light").

It seems that the first version is not only more thorough but also follows the logic of the very culture of belief that was emerging at this time. To say that Fantasia gave Homer the ideas for his poetic masterpieces is similar to the idea that the Muse would have done this. The muse, as the first example of metalanguage, in principle, receives an important place in the artistic language of the ancient Greeks. Here, the inner "other" begins to make its way outside, albeit unconsciously.

Homer's Odyssey, translated by William Cullen Bryant, begins with these words:

> Tell me, O Muse, of that sagacious man
> Who, having overthrown the sacred town

Of Ilium, wandered far and visited

The capitals of many nations, learned

The customs of their dwellers, and endured . . .

The reference to the muse here is not an artistic invention of the translator, and it is also present in the original:

Ἄνδρα μοι ἔννεπε, Μοῦσα, πολύτροπον, ὅς μάλα πολλά . . .

"Μοῦσα" is a muse. Since ancient times, authors have considered the muse as a source of inspiration. According to the ideas of the ancient Greeks, it was thanks to the muses (the kiss of the Muse) and not because of a mental process that the authors had outstanding ideas. The Muses are the deities of the companion of the patron god of the arts, Apollo. But as we apparently understand today, the muses do not exist and do not reside at the very top of Mount Helicon.

The source of an artist's inspiration lies within the artist himself, and this source is called a "special view" of things. Its mental basis should be the view from the "other", which is accessible to the author himself. In other words, we believe that those very depths of the subconscious, to which miraculous insights are usually attributed, are nothing more than the repressed personality of the "other."

In the mechanism of persuasion, this personality is not yet recognised by the author, and therefore, the culture of persuasion at the first stage turns out to be full of references to some imaginary third parties. The unconscious "other" is trying to return to consciousness and, according to Freud's metaphor, knocks on all doors. We convince the "other" not to knock, but this is only a temporary respite. Having entered the path of persuasion, we now constantly must convince each other, as if not noticing that we are convincing ourselves.

The culture of persuasion is a dialogue with oneself that the author conducts with another object. In another person, in another thing. Greek sculpture tries to depict not what a person is like but how the author of the sculpture unconsciously wants to see himself. The Roman speaker, before convincing his audience, first convinces himself and then invites the audience to take their place, to see the situation with their own eyes. The author of a culture of persuasion shares with the audience what he believes the world should be. This is what makes this art classic because there is still no conscious splitting of the subject and, consequently, a protest against its nature.

It becomes important for power in this culture to have legitimacy. Power becomes a form of dialogue with people. Both in a republic, in a monarchy, and in mixed types of government, a format of belief in the legitimacy of power appears. Now, it is not enough to be the strongest, but it is important to convince everyone that you are the ruler. The monarch under autocracy receives his power from God, and the Council of the Republic receives it by being elected by other people. Power is no longer part of the personality of the ruler, as in a culture of coercion, but can be given and alienated.

The legitimacy of power in a culture of persuasion comes from the "other." But this "other" is not yet recognised as one's own part. They believe in this power without noticing the actual situation behind it. The rationality of power is repressed into the unconscious realm, and it was only possible to comprehend this many years later through the work of Thomas Hobbes. Aristotle, in his *Politics,* says that the state is a natural formation. But, of course, only society is natural, and the state is an element of culture. However, the ancient Greek did not yet realise this, since his "other" seemed to him something external. God or justice. The fact that he himself is both one and the other did not immediately become clear.

The next step in the development of the psyche and culture at the same time was connected precisely with this awareness of one's "other" with the ability to discern objective motives in one's own and others' actions. Today we see much more than what is commonly talked about. We see the motives for people's actions, often very mundane and simple, and we just as often forgive them because we know that we, ourselves, very often have the same motives. For a very long time, such a train of thought was not accessible to man since his "other" seemed to him to be something external.

Awareness of the "other" as part of one's personality was very difficult and, even by historical standards, took a long time. St. Augustine began to take approaches to this bastion when he introduced the concept of "original sin" in his treatise 396 *On Various Questions to Simplicity*. Augustine wrote that human nature is damaged and that this damage can only be healed by faith. All great religions are, in one way or another, part of a culture of belief, and with the statement that only faith can heal the soul, Augustine addresses precisely this culture. However, substantively, he already outlines the question that some sinful or natural part of the personality is part of the person himself.

We pointed out above that a frequent theme among Christian thinkers was the sinful nature of man. Pope Innocent III wrote the treatise *De miseria conditionis humanae* (Towards human poverty) (1194-1195). This is a very dark work in which the author describes the depth of sinfulness of human nature. On the one hand, the pathos of this work is in the humiliation of the sinful nature of man, but, on the other hand, in this humiliation, which in the works of various authors stretched over centuries, the idea is already maturing that this sinfulness has nothing to do with the devil (the "other"). After all, it's just us.

In the greatest poets of the High Middle Ages and the Renaissance, Dante, Chaucer, and Boccaccio, we already find images of non-mythological but completely life-like characters who combine the low and the high. For example, the English poet Geoffrey Chaucer (1343-1400) in *The Canterbury Tales* cites the story of The Wife of Bath's Prologue and Tale.

He presents this woman as a cynical and calculating person who married five times. Although this is just a "low" woman, Chaucer devotes twice as much space to her story in his book than to the story about a knight from the time of King Arthur. And today (in the 21st century) the weaver from the city of Bath appears to us in a completely new light as almost the first feminist in literature.

Chaucer writes about common people for common people. And the weaver from the city of Bath is a more interesting person for him than the noble knight. The old values created in the era of persuasion no longer work as effectively. The heroes of this time are people who live by the values of nature and society that are understandable to everyone.

The "other", which has been imprisoned in the realm of the unconscious for so long, is no longer a subject for doubt and shame. This means that the culture of persuasion here is developing into something fundamentally new. The "other" is now realised, and the balance of efficiency and justice is again upset. It is difficult for a person to realise his true nature. This imbalance must be compensated. As a result, two cultures consistently emerge: the culture of enlightenment (education) and the culture of inspiration (transformation).

The first of them, the culture of education, equalised the "other" in mental rights, recognising his nature, but imposing strict restrictions on him in his manifestations. This culture restored balance due to the fact that the systematic study of the "other" became an independent cult – the cult of knowledge. This cult means reflection, the study of one's inclinations and qualities, the identification of "bad qualities," and the subsequent creation of conditions in which these "bad" qualities should not manifest themselves. The idea is quite in the spirit of Plato's *Republic* but taken to a new level.

In the culture of education, part of the psyche becomes publicly available and the subject of open discussion without unnecessary, inappropriate restrictions. This also gives rise to the permissibility of the vulgar "low" style, which, however, consists only in the recognition of the "other" and ironic retribution for him.

The second, the culture of inspiration, is a reaction to the culture of education, and outwardly it is expressed in the idea that: "Human nature is after all good." This culture can no longer push the "other" back into the unconscious, but it is built on the fact that the "other" is only a small and insignificant personality in the structure of the psyche. The culture of inspiration arises at the psychic level as overcompensation of the "other," which could be called the "super-ego." The "super-ego" is a leader personality in the structure of the psyche, which arises in a person as a reaction to the awareness of the "other."

Let us consider several key points in the history of how these cultures were established. And how they affected the current situation. So, the culture of education is to reject meta-text. It is no longer the Muse, other deities, or other third parties who are to blame for what is happening. The "other" is realised and scrupulously studied under the microscope of the artist and ruler, striking us with its intricacy.

In 1513, Nicolo Machiavelli wrote *The Prince*, in which he takes it for granted that man is not perfect. It reveals what the ruler's motives should be in order to achieve the result necessary for the state.

The German philosopher Rudolf Gocklenius (1547-1628) uses the concept of "ontology" to describe a certain doctrine about an object taken in abstraction from the subject and his life activity. The history of modern psychology, and essentially a psychology that recognises the status of the psyche, begins with Goclenius. It is fundamentally important for us here that psychology, which seeks to objectively study human motives in its meaning, is a product of the culture of education. After all, it is psychology that allows us to understand human weaknesses and distinguish between those that are inevitably inherent in everyone and are the norm and those that deviate from the social norm. Moreover, it is in psychology that "evil" becomes a disease and is studied scientifically.

The "other" comes out thoroughly in works about the famous Don Juan. His first artistic image appears in 1630 by Tirso de Molina in the play *The Mischief of Seville, or the Stone Guest*. In this work, perhaps for the first time, negative traits of human character are popularised. Yes, Don Juan mocks and humiliates his victims, but the author does not call him a scoundrel. He calls him a mischief-maker. This word, which is more suitable for a child than an adult, shows Tirso de Molina's attitude towards this phenomenon.

A well-read reader will probably be indignant and ask: "But, excuse me, what about the Metamorphoses of Apuleius?" Wasn't it there, in the famous "Golden Donkey", that back in the 2nd century, the adventures of the dissolute noble Roman youth Lucius were first described? Lucius was passionate about women and, worse, witchcraft. Isn't this where our "sympathy" for the "other" in us originates? No, not here.

Unlike the hero of the early Renaissance, Lucius sought God. Wasting his youth was not interesting to him, and as a result, he renounced his former vicious life. Apuleius is studied and quoted by Augustine himself, realising that this literature, although provocative, is in many ways instructive for young people seeking worldly pleasures. She gives them a model of punishment for sins before they occur. But in addition to the denouement (morality), there is also a fundamental difference in the motives of the heroes. Lucius's pleasures cannot be called mischief; he simply strives to please and satisfy his needs. Lucius, at first, does not understand that he is doing something wrong.

Don Juan, however, is being mischievous. That is, he is aware of the meanness of his deeds, and it is this meanness that amuses him. In the future, we know that this image will rise to become the Byronic hero in the 19th century. A whole galaxy of images of "evil" and smart young intellectuals will appear on the stage of world culture one after another. This is actually Byron's Child Harold, this is John Polidori's "Vampire", and, among other things, Pushkin's Onegin and Lermontov's Chatsky.

The image of an evil mind is still very popular in art. But we are not talking about evil in the ordinary sense because the most important thing here is not the ethical, but the intellectual component. Reason is above good and evil, so the heroes of the culture of enlightenment (that is, a culture that is aware of the "other") often look like evil. This is, say, Faust (1808) and Woland (1966). This "other" may be present metaphorically, but we already understand perfectly well what we are talking about here.

The understanding that the "other" must be curbed by systemic means generally comes in the mid-17th century. In 1651, Thomas Hobbes, in *Leviathan*, openly stated that without external control, a person is capable of the most terrible crimes. Having experienced numerous wars and upheavals, Hobbes knew firsthand how bad human nature can be despite any cultural achievements of a bygone era of belief. Hobbes does not propose to close our eyes to the "other" in us but, on the contrary, to recognise it in all its ugliness and create an enlightened society of those who know about themselves. And this society, in his opinion, for the purposes of community life, must implement an agreement in which it will be openly stated that the "other" is now outside of society.

It is very important to note that this position – we know the "other," and he is now outside of society – is a conservative position in the broadest sense. It is based on the progress of thinking, on the need to educate the broad masses of the population, but culturally it creates a certain ideological dead end. The institution of education becomes public knowledge, that is, knowledge in a certain sense of the achievements of modern science at some level. Why science? Because the study of "the other," as we noted above, is a subject of objective study.

The encyclopaedia becomes a symbol of the culture of enlightenment as a kind of body of accumulated knowledge. But is knowledge such an immutable value? At the psychic level, knowledge of the "other" or objective knowledge creates an unstable balance. As an example, it can be noted that even the most convinced of the objectivity of the laws of nature, scientists quite often, already in adulthood, begin to believe in the supernatural, that is, in God. And the most inveterate atheists remain so rather because their atheism is a type of faith (value) or ideal.

In other words, knowledge itself is a potentially unstable ideological basis, although it is the most reliable. In addition, the development of a knowledge system is a specific activity that requires both motivation and preparation. Therefore, we believe that a knowledge-based worldview is, by modern standards, a conservative worldview.

The exposed "other," which in one version of psychic science is referred to as "it," generates in the psyche the need for a "super-ego." This is an option that allows you to avoid "escaping from yourself" and from the world. The "super-ego" or the internal image of an ethical, aesthetic, political or any other ideal arises both as a cultural concept and as a mental phenomenon. The appeal to the "super-ego" of people can be called inspiration in our special sense. Inspiration here is understood as one of the forms of motivation for efficiency.

Inspiration as such a special form of activity appears based on enlightenment, but from the very first steps one must dissociate itself from it. What is later institutionalised in the special mission of the party system, at the very beginning of the journey, looks like a daring attempt to humiliate the role of the "other" in oneself. And this attempt relates to the personality of the French philosopher Jean-Jacques Rousseau.

Believing that man is bad, Hobbes proceeds from the need for state control over him. And, strange as it may seem, it is precisely from this seemingly overly conservative view that English liberalism originates. He, in combination with the ideas of John Locke, ends up in the USA. Rousseau is often blamed for the fact that his ideas about the inherent goodness of people led to revolution in France, and then throughout the world.

It's not that simple, of course. Rousseau expressed the ideas that were in the air at that time: ordinary people are no worse than aristocrats and, in many respects, perhaps even more worthy since they can be kinder, more hardworking, and so on. Rousseau says that it is not ordinary people who need to be enlightened, but the aristocracy who needs to learn from the commoners their "craft", their "life", in which there is a place for "dirty" work, caring for themselves and others.

Louis XVI was the last monarch of the old order in France, and the exponent of that absurd greatness with which coercion (the aristocracy) and persuasion (the clergy) reached their apogee. It was these institutions that could no longer keep a weak and short-sighted monarch on the throne. The personality of the monarch played an important role in the moment precisely because the personality of the "other" was at that moment the subject of close attention of everyone without exception. And the path of enlightenment led to the constitution, and the path of inspiration (transformation) to revolution. At the very beginning, both paths were very similar, but very soon, it became clear how much they differed.

Class society began to die out, but how then to maintain a balance of efficiency and justice? A new form of division was emerging – party division. Party, in fact, comes to replace class.

Leftists appear in place of the oppressed class, and conservatives replace the ruling class. The right expresses the interests of the bourgeoisie. Here, the culture of enlightenment finally fits into the conservative agenda because it is not ready to recognise the "super-ego". The culture of enlightenment does not appeal to ideals but only to dry reason. But, my God, how unbearable it is, how difficult it is to remain cold and objective. This culture seems to have survived as a dominant way of thinking in Europe to this day, mainly only in England. And then in a highly transformed form.

Separated from the actual situation, the "super-ego" is another round of culture, a path of inspiration by ideals. The moment the "other" in us comes out and ceases to be a sacred secret, the power of the old institutions weakens. But a new form of utopia is emerging – a new clergy under the name of *visionaryism*.

The peculiarity of the "super-ego" is that it is empty in content. This is a clean slate for culture. It is no coincidence that Rousseau writes that in a society built on a social contract, the will of each individual person is alienated and transferred to the exponent of the general will. That is, there is a delegation of responsibility to the very top – to the party super-ego. This is exactly what is happening in many countries. New party leaders become carriers of the image of the "super-ego" as if taken outside the psyche.

The USA is no exception here. The Revolutionary War created the "American Dream" society, a society of superheroes, super achievements, and high ideals. And ideals, as you know, are called ideals because they are unattainable. Hollywood is called the "dream factory," and this factory competed well in an ideological sense with similar factories of totalitarian regimes around the world. And this competition would not be in favour of Hollywood if it did not have its own ideals, competing with the ideals of these totalitarian regimes.

We will not repeat ourselves about totalitarian regimes; we will only note that Rousseau's ideas turned out to be close to many people. In Russia, for example, this line was continued by Tolstoy, who put a lot of effort into showing the advantages of the common people, tried to learn from these people and see the best. It is not surprising that his work turned out to be so to the liking of the new Soviet government in Russia.

The "super-ego" was also reflected in unreal art, which deliberately distorted reality. Not because it could not exactly repeat the natural original but because behind the natural original, it always saw "another." Repeating reality has become taboo for the art of inspiration. It left this boring mission to the art of education.

So, the "super-ego", which appeared as a psycho-social phenomenon, entered the arena of public life and changed it completely. Today, in all corners of the world, we have societies that, to varying degrees, continue to combine all four forms of motivation. This is coercion by the state, persuasion by the clergy and art, enlightenment by the forces of science and education, and inspiration through the institutions of visionaries and political parties.

In this sense, we all carry within us a heap of cultural products – from traditions and values to knowledge and ideals. And this could continue indefinitely. A culture can make a mixture of these ingredients indefinitely. However, not so long ago, we all witnessed how a completely new mechanism for accumulating useful changes appeared in the arsenal of culture. And very soon this mechanism will greatly influence human civilisation.

Success and the Meaning of Life

In the pursuit of success and justice within their group, people are divided into those who believe that it is necessary to support the weak and those who would prefer to support the strong. Why is this happening? Why is there this difference in positions? It would be more accurate even to ask what the hope is of those who support the weak and what the hope is of those who support the strong.

If we stop at this formulation of these positions, we will never understand what is going on here. After all, supporting the weak is simply mercy, and supporting the strong is just a habit. Neither charity nor habit can be an end in themselves. The first is a natural tribute to our humanism, and the second is an equally understandable tribute to tradition. But what is this strange dispute between humanists and traditionalists? Are humanism and traditionalism mutually exclusive? Of course not. The situation is different.

The secret to this debate is that people have different views on achieving group success. Those who support the weak, that is, the Left, are much more pragmatic than is commonly believed. Behind their support lies the hope that those who need help today are our growing future – the "children" of human society. And children need support. And not helping them means depriving yourself of the future. In this sense, it is understandable why, at the very beginning, the left only supported the emerging working class, but now it is of little interest to them. The future is changing. The Industrial Revolution is already in the distant past, and today, we need to think about new hidden resources of humanity.

There is a second, no less important motive. The desire to make the weak stronger is not simply charity but a desire to expand the circle of those who can fully participate in the common cause, making their contribution and bearing responsibility. By depriving women of civil rights in the distant past, for example, men not only limited their rights but also deprived themselves of help from women, say, in politics. Modern practice shows that women can be very successful as politicians.

After all, what is politics, in essence, if it is practised in good faith? It is the desire for group success in the face of limited resources. But, unfortunately, no one knows the secret of success. And there is no reason to believe that, for example, a white man will cope with this task any better than, say, a black woman. The components of success are will, openness to change, creativity, and so on. Therefore, it is reasonable to hope that involving people in political processes who have not yet participated in it can yield good results. The main thing is that political "newcomers" should be strong in spirit.

The Right sees success completely differently, although the result is the same for everyone. By supporting the powerful, they vote for what has traditionally worked in the past rather than relying on "new" leadership and "new" participants in the economy. However, since we are usually talking about achieving success in the future, the Right does not choose those who have already achieved success (although this would be an ideal choice) but rather those who fit the traditional signs of success in their image. These are necessarily people who have already achieved something in economic activity. These are those who are ready to make unpopular decisions – to cut the Gordian knot. These are people whose rhetoric resembles a doctor's speech when it's time to start treatment and not the inspiring speeches of a "new" leader.

In summary, the main difference between those who support the weak and those who support the strong lies in different understandings of what will lead to group success. And only in second place is the question of what is fairer. It is foolish to think about justice in a group whose members have died of starvation. But is it possible to unambiguously answer the question of which of the two strategies is better? Is it possible to determine what will lead to success in advance?

Let's consider this example. After a disaster, a group of people find themselves in a forest without contact with the world. They need to survive until help arrives and learn to get food in the wild. The group consists of several men, several women and some children. At the same time, during the disaster, a certain amount of food was preserved, and it must be used rationally. In this case, two strategies are possible.

The first strategy is to distribute the food evenly, giving everyone the same chance of survival. After all, it is necessary to preserve the entire group, including the weak and sick.

Moreover, any member of the group can come up with the idea of salvation. In an extreme version of this strategy, more resources may be given to the weak since it is more difficult for them to endure deprivation. In this case, a larger number of people will remain alive when rescuers arrive or by the time someone in the group figures out how to save the group.

The second strategy is to give more food to the most able-bodied men, who potentially need more energy to go hunting and get food or make a long foray and find salvation. The first strategy will be left-wing, and the second right-wing. But which one will lead the group to success? The problem is that this is unknown and depends on many unknown factors that were not initially specified in the problem. It is unknown how long the wait for rescuers will last, whether there are populated areas nearby within walking distance, whether there is enough food in the forest for this group that can be obtained, and whether someone from the group will come up with an idea on how to find salvation. It is also unknown whether the men will be able to catch a conditional wild boar or make a successful foray, even if they are fed better than the rest.

In this sense, the classic depiction of children marooned on a desert island from the novel *Lord of the Flies* is indicative. More right-minded children did not want to play "kindergarten" with discipline and stupid rules because they lost faith that help would come. They chose the path of hunting, survival at any cost, and demanded that the second group give them the knife (their only weapon). They did not want to support the weak in the hope that success would come. According to the traditional laws of the wild, the strongest would win. In the book, however, this group is shown as an unsuccessful example since help does come, but the children have already committed a number of unforgivable crimes. But what if help had not come?

It seems that justice should be clear in any situation, but it still depends on a lot, including such an unpleasant factor as luck. The psychology of individuals in a group, especially leaders, also plays a key role. With any of the chosen strategies, unsuccessful leadership can lead to the death of the group. An outwardly more just leader may be less successful and vice versa.

The unpredictability of the future makes success mysterious. If success is mysterious, then choosing a strategy ceases to be just an intellectual process but also includes intuition. This is an intuition about how strengthening the weak and striving for equality is more effective than strengthening the strong and striving for inequality. We can say that supporting the weak is not only the hope that the weak can become stronger and help others in the common cause but also the intuition that universal equality is generally strategically more successful. We can also say that supporting the strong is not only the hope that the traditionally strong will be able to cope with complex challenges better than others but also the intuition that universal equality, in general, is strategically less successful.

These two positions can be expressed by maxims. The first one would sound like this. It is more successful to give everyone the same opportunity to solve common problems. And the second – so. It is more successful to provide opportunities in solving common problems to those who are already more successful. If we rephrase these maxims, the first position is that success is completely unpredictable, and those less successful today have an equal chance of being more successful in the future. The second position is that success is partly predictable and rather accompanies those who have already been successful.

This is why people on the Right can sometimes look at people on the Left and think they are being too romantic. In fact, this is not romanticism but a progressive approach to assessing the likelihood of success. Whereas on the Right, this approach is rather conservative. Moreover, "progressive" means not focused on past experience but focused on the method of revaluing existing human resources. It can be said that the Right is more inclined than the Left to consider the world and man as knowable and, therefore, are more inclined to know the "meaning of life." Whereas the Left is less inclined to know the "meaning of life" and believes that it may be completely incomprehensible. And so, they act based on this position. If we do not know the meaning of life, we should value not the ends but the means.

Ultimately, the two positions come from different assessments of what a person is and whether a person is special for some reason. If a person is special and has a special mission in the world, then he is valuable in himself without any additional "meaning of life." If a person is not special, then some goal is a priority.

The only thing that possibly makes a person special, different from the world around us, is our consciousness. But the trouble is, we don't know what consciousness is. It turns out that this is the reason for all the disputes, discussions, and wars. Because of this, there are Left and Right. Having solved the problem of consciousness, we will probably understand who we are, whether we are special, and whether any other goal other than humanism is possible.

Bibliography

Ake Sjoberg and E. Bergmann (trans.), *The Collection of the Sumerian Temple Hymns*, J.J. Augustin, 1969.

Baars B.J., *A Cognitive Theory of Consciousness*, New York: Cambridge University Press, 1988, [Kindle edition, 2012].

Chalmers, D. J., *The Conscious Mind: In Search of a Fundamental Theory*, New York: Oxford University Press, 1996.

Chomsky, N., *Syntactic Structures*, The Hauge: Mount, 1957.

Crick F., Koch C., (1990), 'Towards a Neurobiological Theory of Consciousness'. *Sem. Neurosci.* 2, pp. 263-275.

D. Chalmers, *Reality+ Visual Worlds and the Problems of Philosophy*, London: Penguin Books, 2022.

Damasio, A.R. (1989), 'Time-locked Multiregional Retroactivation: A System-Level Proposal for the Neural Substrates of Recall and Recognition', *Cognition*, 39, pp. 25-62.

Dennett, D. C. (1978). 'Toward a Cognitive Theory of Consciousness'. In D. C. Dennett (Ed.), *Brainstorms* (pp. 149-73), New York: Bradford Books.

Edelman, G. M., and Tononi G., *A Universe of Consciousness: How Matter Becomes Imagination*, New York: Basic Books Inc, 2000.

Freud, Sigmund, in Gay, Peter (ed.), *The Freud Reader* (1st ed.), New York: W.W. Norton, 1989 (pp. 664–665).

G. Tononi, O. Sporns, and G. Edelman (1992). 'Re-entry and the problem of integrating multiple cortical areas: Simulation of dynamic integration in the visual system', *Cerebral Cortex*, 2:310-35.

John Searle, *The Rediscovery of the Mind*, Cambridge Mass: MIT Press, 1992.

Jung, C. G., *The Archetypes and the Collective Unconscious*, Princeton, NJ: Princeton University Press, 1980.

Libet, B. (1978) 'Neuronal vs. Subjective Timing, for a Conscious Sensory Experience', in: *Cerebral. Correlates of Conscious Experience.*

Marx, K., *Das Kapital, a Critique of Political Economy*, Chicago: H. Regnery, 1959.

Matthew Botvinick & Jonathan Cohen, (1998), 'Rubber Hands "Feel" Touch that Eyes See', *Nature* volume 391, (p. 756).

Merleau-Ponty, M. (1964). *The Primacy of Perception* (J. M. Edie, Ed. And Trans.), Evanston: University of Illinois Press.

Meshcheryakov, A.I., *Awakening to Life* (K. Judelson, Trans.), Moscow: Progress, 1979

Nagel, T. (1974), 'What is it Like to Be a Bat?' In N. Block (Ed.), *Readings in Philosophy of Science* (pp. 159-70). Cambridge, MA: Harvard University Press.

Oittinen, V. K., & Levant, A. (2013), 'Dialectics of the Ideal: Evald Ilyenkov and Creative Soviet Marxism', *Historical Materialism* Book Series; Vol. 60.

Penrose, R. (1987), 'Quantum Physics and Conscious Thought', in B. Hiley and Peat, eds., *Quantum Implications: Essays in Honor of David Bohm*. New York: Methuen.

Putnam, H. (1960), 'Minds and Machines', In S. Hook, ed., *Dimensions of Mind*, New York: New York University Press.

R. Descartes (1647). 'The Passions of the Soul', in J. Cottingham, R. Stoothoff, and D. Murdoch, eds., *The Philosophical Writings of Descartes*, vol. 1. Cambridge: Cambridge University Press, 1985.

Ryle, G., *The Concept of Mind*. London: Hutchinson, 1949.

Simondon, G. (1958), *Du mode d'existence des objets techniques*, Paris: Aubier, 1989

Simondon, G. (1980), 'On the Mode of Existence of Technical Objects' (Part 1), Ninian Mellamphy, (trans.) (with a Preface by John Hart), University of Western Ontario.

The Code of Hammurabi, Translated by L. W. King. THE AVALON PROJECT Documents in Law, History and Diplomacy. http://avalon.law.yale.edu/ancient/hamframe.asp

Vygotsky L. S., *Collected Works: In 6 Volumes*. T.3 Problems of mental development / Ed. A.M. Matyushkina. - M.: *Pedagogy*, 1983, - 368 p., Ill.

Vygotsky, L. S., *Thought and Language*, Cambridge MA, MIT Press, 1962.

Wiener, N., *Cybernetics: On control and Communication in the Animal and the Machine* (2nd Ed.), Cambridge, MA: MIT Press, 1961.

Wundt, W. (1912-1973). *An Introduction of Psychology*, London: G. Allen, Reprint, New York: Arno Press.

Yasnitsky, A. (Ed.), *A History of Marxist Psychology: The Golden Age of Soviet Science*, Routledge/Taylor & Francis Group.

Printed in Dunstable, United Kingdom